ELSEVIER'S
DICTIONARY OF
HERPETOLOGICAL
AND
RELATED
TERMINOLOGY

ELSEVIER'S DICTIONARY OF HERPETOLOGICAL AND RELATED TERMINOLOGY

compiled by

D.C. WAREHAM
Bournemouth, England

2005

ELSEVIER

Amsterdam – Boston – Heidelberg – London – New York – Oxford
Paris – San Diego – San Francisco – Singapore – Sydney – Tokyo

ELSEVIER B.V.	ELSEVIER Inc.	ELSEVIER Ltd	ELSEVIER Ltd
Radarweg 29	525 B Street, Suite 1900	The Boulevard, Langford Lane	84 Theobalds Road
P.O. Box 211	San Diego, CA 92101-4495	Kidlington, Oxford OX5 1GB	London WC1X 8RR
1000 AE Amsterdam	USA	UK	UK
The Netherlands			

First edition 2005

Library of Congress Cataloging in Publication Data
A catalog record is available from the Library of Congress.

British Library Cataloguing in Publication Data
Elsevier's dictionary of herpetological and related
  terminology
  1.Herpetology - Dictionaries 2.Amphibians - Dictionaries
  3.Reptiles - Dictionaries
  I.Wareham, David C. II.Dictionary of herpetological and
  Related terminology
  597.9'03

ISBN: 0-444-51863-0

Printed and bound by CPI Antony Rowe, Eastbourne

CONTENTS

ABOUT THE AUTHOR

David C. Wareham has been studying and working with captive and wild reptiles and amphibians for over forty years. Until 1955 he was Curator of the prestigious Cannon Aquarium and Vivarium at the Manchester University Museum in northwest England, where he received much acclaim for his success at breeding several so called "difficult" species.

Through his practical involvement, on an almost daily basis, with a huge variety of reptiles and amphibians from around the world, he has acquired an extensive knowledge and, in between his field studies of wild populations, is now sharing it with others in his articles and books on the husbandry, natural history, and conservation of these fascinating animals.

FOREWORD

Herpetology is like any other branch of science, it is littered with specialized words and complex phrases. Some are quaintly archaic and rarely used, while others have been coined much more recently and relate directly to recent biochemical, genetic or microbiological developments and advances in our understanding of the natural world. Herpetology incorporates everything from Linnaeus's earliest taxonomic classifications and Wallace's ideas concerning biogeography, to ELISA comparisons of blood serum and snake venoms and CITES-initiated protection for endangered species. Herpetology is a broad church, even swallowing its nearest relative, the science of Batrachology, as amphibians as well as reptiles (extant and extinct) fell under its vast umbrella.

It is probably the language of science that persons new to its cloisters find most daunting. Some words are difficult to pronounce, others are confusingly similar in appearance i.e. 'parotidgland' and 'parotoidgland' (what a difference an 'o' makes!). Specialized terms and phrases seem, to the outsider, elitist and scary. People are afraid to use them in case they pronounce them incorrectly, use them improperly or alienate their audience. I know that book publishers dread a glossary that runs to more than a single page, and television producers fear a profusion of 'technical words' issuing from the lips of the presenter or contributor, believing such speech will cause the viewers to change channel in droves. The common question is: "Do you have to say that? Can you not find a simpler word?". Should we dumb-down 'parthenogenesis' and use 'virgin birth' instead, bearing in mind it is more probably "virgin oviposition", or is it still too technical?

Science is governed by specialized words and phrases, just as law, politics, and economics have their own languages, much of them incomprehensible to the outsider. Notice, I say science is 'governed', science is not hampered by these words. Only the person who does not understand the language of science is hampered by these words. And that is where a book such as this is so important.

A dictionary of herpetology is not a new idea. I have a 1964 copy of a James A. Peters publication by that very title and also David C. Wareham's 1993 *Reptile and Amphibian Keepers Dictionary*, and there have been and will be others. The point is that dictionaries have to grow, evolve and advance with the science they purport to explain as new terminology is developed to keep pace with understanding and research. In preparing the *Elsevier's Dictionary of Herpetological and Related Terminology* David C. Wareham has brought together both the enduring and the latest terminology of herpetology in one easy to follow publication, drawing on all its specializations from systematics to snakebite. In so doing he is providing a valuable resource to everyone with an interest in herpetology-hobbyist, student, author, professor, as a quick source of definitions. I am confident you will wonder how you ever managed without this dictionary before.

Mark O'Shea
Herpetologist with field experience of six continents.

PREFACE

This book has been based on my now out-of-print *Reptile and amphibian keepers dictionary: an A to Z of herpetology* which was published by Blandford Press in 1993. This new, revised edition has been updated and now includes over 3100 terms and their definitions, more than 1000 than the original.

The dictionary is a compilation of clear, concise and informative definitions of the characteristic vocabulary commonly encountered and used by herpetologists when discussing, reading, or writing about reptiles and amphibians. It is intended for all those who have an interest in these animals, from the amateur hobbyist who may find himself faced with what can be a rather intimidating scientific term or technical expression, to the trained zoologist who may sometimes have doubts over the exact meaning of a particular term. It will, it is hoped, not only be a useful source of reference to all who are either actively or passively involved in some aspects of herpetology whether they be keepers, curators, breeders, researchers, teachers or students, but an interesting read as well.

The entries, which are fully cross-referenced, include all the basic technical terms relating to the external features of reptiles and amphibians (e.g. femoral pore, parietal scale, supralabial gland, etc.), the herpetological families (e.g., Bufonidae, Iguanidae, Viperidae, etc.), biological processes (e.g., aestivation, spermatogenesis, thermoregulation, etc.), selected biographies (e.g., Bellairs, FitzSimons, Smith, etc.), herpetological jargon (e.g., hoody, in-egg, milked up, etc.), and many other terms and expressions of relevance to herpetologists, from the fields of anatomy, ecology, zoogeography, toxicology, veterinary science, animal behaviour, and husbandry, etc.

The headwords are arranged in alphabetical order and wherever a particular term contains more than one word the order of those words remains the same. For example, 'metatarsal tubercle' appears under M and not under T as 'tubercle, metatarsal'. Words appearing in upper case letters within the definitions are themselves defined elsewhere in their correct alphabetical position. Synonyms, where appropriate, are included immediately after the headword. The dictionary also includes abbreviations, acronyms, and symbols relevant to the study and conservation of reptiles and amphibians.

It has been impossible to include every single herpetological term ever devised but it does however contain the majority of those currently in use, together with some others which are perhaps not quite so familiar. As with any compilation of this kind, this work has relied almost totally upon the work of others who have contributed to the science of herpetology. They are far too numerous to mention but without them this book would not have been possible. Thanks too are due to all the many friends and colleagues who have offered much help, advice and encouragement during this new book's progress and who have suggested terms for possible inclusion. I am indebted to them all and, in particular to Linda Versteeg, publishing editor at Elsevier, for her help and suggestions regarding the text, and my friend and fellow herpetologist, Mark O'Shea, who has supplied me with many terms especially those relevant to venomous snakes and the effects of their bites.

David C. Wareham
Dorset, England, 2004

A

A'ASIA Abbreviation for Australasia.

AAZPA Acronym denoting the American Association of Zoological Parks and Aquariums.

ABBREVIATION A shortened contracted form of a word, phrase, or title used in place of the whole. In zoological literature the genus name referred to in the BINOMIAL NAME of a species, when quoted more than once in the same work, is often abbreviated to just one letter followed by a full stop. For example, in the case of the Northern viper, the full scientific name *Vipera berus* is used the first time it is mentioned, with the abbreviated *V. berus* used subsequently. The same method is used for both the generic and specific names referred to in the TRINOMIAL NAME of a subspecies, thus, *Pantherophis (Elaphe) obsoleta quadrivittata* the yellow rat snake (a subspecies of the American rat snake *Pantherophis obsoleta*) is used in full on the first occasion and *P. o. quadrivittata* subsequently.

ABDOMEN 1. The region of the body, usually the hindmost section, which contains the reproductive system and the major part of the digestive system; the belly. **2.** The complete undersurface which, in limbed amphibians and reptiles, is the area lying between the fore- and hind legs and, in limbless forms, between the throat and the CLOACA; the VENTER.

ABDOMINAL Pertaining to the region of the ABDOMEN, also a scale or LAMINA on the undersurface of a reptile which, in lizards, is any scale lying on the ABDOMEN and, in snakes, is any one of the enlarged VENTRAL scales. In chelonians, the abdominal is either of the pair of laminae lying third from the rear on the PLASTRON.

ABDOMINAL GLAND 1. An enlarged gland situated inside the CLOACA in male urodeles which, in certain species, may serve to stimulate the female during courtship. **2.** In the casque-headed skinks (*Tribolonotus*), any one of a number of glands, each located beneath an abdominal scale, and a feature unique among lizards.

ABDUCTOR MUSCLE Any muscle the contraction of which acts in drawing a structure, such as a limb for example, away from the centre line of the body.

ABERRANT Abnormal or diverging from the usual or normal form. Among reptiles and amphibians, an aberrant individual exhibits, in one way or another, a feature or features uncharacteristic of its species. A snake that shows the characteristics of albinism, or one that has a pattern defect where spots are replaced by stripes, are examples of aberrant specimens. Several similarly oddly-marked or patterned individuals within a species usually suggests

INFRASUBSPECIFIC status.

ABIOTIC *See* ABIOTIC FACTOR.

ABIOTIC FACTOR Any feature which is ecological or non-biological but which still plays an important role in an organism's environment such as temperature, humidity, light intensity, soil pH, or forest fire, for example. Abiotic factors are in essence the result of natural changes in an environment's stability.

ABIOTIC STRESS Any ABIOTIC FACTOR that has a harmful effect on organisms in a particular environment. Pollution resulting in acid rain for example causes abiotic stress by reducing a species capability to survive in the environment.

ABORAL Situated away from or opposite the mouth.

ABRUPTION A breaking off of a part or parts from a mass. Sometimes used in reference to tail shedding (AUTOTOMY) in certain lizard species for example.

ABRUPT SPECIATION In accordance with Darwin's work on NATURAL SELECTION and ADAPTIVE RADIATION, the development of a new species over a comparatively short period of time.

ABSCESS A localized gathering of pus, often caused by bacteria and occurring on various parts of the body. In tortoises, abscesses are usually found around the ears, and in small lizards they generally occur in the area of the CLOACA. In other reptiles, abscesses can occur almost anywhere, but particularly on the head in snakes.

ABSENTEEISM A form of behaviour seen in crocodilians and many other species in which parent animals leave their offspring to support themselves. Although either or both parents may stay close by or return intermittently to check on or protect their exposed offspring, their young's independence is encouraged so that they will adjust rapidly to their harsh environment.

ABUNDANCE In reptile and amphibian ecology, abundance refers to the density of specimens, less commonly of species, within a specified area.

ABUNDISMUS A variety of MELANISM in which a particular specimen displays one or more variations to its basic species pattern, e.g., an increase in the amount of black markings in areas which are on the whole non-black, whilst not being entirely black, as in true melanism for instance.

ABYSSAL Of, in or pertaining to that part of the ocean which has a depth below 2000 m. *Compare* BATHYAL; and NERITIC.

ACANTHOCEPHALA (Sing.) **ACANTHOCEPHALUS** Spiny-headed worms. A PHYLUM of parasitic worms that, with a hook-equipped proboscis, securely fasten themselves in the intestines of a host. Adults of the GENUS *Acanthocephalus* are sometimes found in water snakes, and those of the genus *Neoechinorhynchus* are frequently found in tortoises. The spines of acanthocephala which inhabit anurans often perforate the stomach and intestines, resulting in their death .

ACANTHOCEPHALAN 1. Belonging or relating to the ACANTHOCEPHALA. **2.** Any one of the wormlike parasitic invertebrates of the phylum Acanthocephala, the adults of which possess spiny proboscises and dwell within the intestines of their vertebrate hosts.

ACANTHOID Having the appearance of a spine; spiny.

ACARIASIS Tick or mite infestation that occurs as ENDOPARASITOSIS but most often as ECTOPARASITOSIS. Newly captured reptiles, particularly snakes, tortoises and varanids, are often infested with ticks, but the most significant ectoparasitosis in reptiles is caused by the blood-sucking mites *Ophionyssus* which occur on snakes. *See* MITE *and* TICK.

ACARID A small arachnid of the order ACARINA such as a MITE or TICK

ACARIDOMATIUM Sometimes found in early literature and relating to the dermal pocket or pouch formation at the rear of the limb insertions of certain lizards. Literally, 'mite house', in reference to the ACARI MITES which sometimes reside within these pockets.

ACARINA An order of tiny arachnids such as the ticks and acari mites. There are around 20,000 known species, many of which are parasites of veterinary and medical importance. *See* MITE *and* TICK.

ACARI MITE *See* ACARINA

ACAUDAL Having no tail. The majority of the Anura for example are acaudal.

ACCESSORY ABDOMINAL PLATE Has been used to describe the foremost row of dorsal scales when distinguishing between a smooth or slightly keeled row of scales where, in certain species, the remaining dorsals are keeled.

ACCESSORY FANG *See* REPLACEMENT FANG.

ACCESSORY PALPEBRAL Sometimes used to describe any one of the scales that lie between the PREOCULAR and the true PALPEBRAL scales on the heads of certain lizards.

ACCLIMATION A term used most commonly in respect of animals maintained in laboratory conditions and in reference to the evolution (development) of physiological adjustment. Acclimation is the response of an animal that allows it to endure a single feature change (e.g. temperature) within its environment. *Compare* ACCLIMATIZATION. *See* ADAPTATION.

ACCLIMATIZATION A reversible reaction that allows animals to withstand and adapt to more than one feature change within their environment (e.g. an increase or decrease in temperature and the resulting abundance or lack of food). Although the reaction is physiological it may effect behaviour (e.g., causing the animal to enter

4

hibernation, or to seek nesting material etc). It also refers, in herpetology, to the way in which a particular natural reaction is diminished or even eradicate over a period of time. For example, crocodilians and large boids which are used in shows and exhibitions can become acclimatized to their 'appearances' and cease to act defensively. The urge to intimidate or take flight can also disappear. *Compare* ACCLIMATION. *See* ADAPTATION.

ACENTRIC Having no body, or centre; used in reference to the vertebrae of amphibians.

ACEPHALOUS Lacking a distinct head. The blind snakes (TYPHLOPIDAE) and the Calabar ground python *Calabaria reinhardtii* could be described as being acephalous.

ACETYLCHOLINE A chemical neurotransmitter produced in the PRESYNAPTIC MEMBRANE that carries a nerve impulse across a SYNAPSE.

ACETYLCHOLINESTERASE An enzyme which breaks down ACETYLCHOLINE at the POSTSYNAPTIC MEMBRANE to prevent continual nerve firing.

ACICULAR Spiky or needle-shaped.

ACINOUS Relating to any GLAND that is subdivided into small cavities or pouches. In male anurans, the NUPTIAL PAD is acinous.

ACOUSTICO-LATERALIS SYSTEM *See* LATERAL LINE SYSTEM.

ACROCHORDIDAE Wart snakes.

Family of the HENOPHIDIA, inhabiting South-East Asia, Indonesia, Papua New Guinea and northern Australia. Three species in two genera.

ACROCHORDOIDEA The filesnakes. A SUPERFAMILY of streamlined, aquatic, live-bearing snakes characterized by flattened, angular heads, movable jaws, and small ventral scales. They lack a vestigial pelvis and are unique amongst snakes in possessing small hair-bearing tubercles the purpose of which is currently unknown. *See also* ANILOIDEA; BOOIDEA; COLUBROIDEA.

ACRODONT Having teeth fused to the top of the upper jaw margin. Acrodont teeth are found in amphibians, the tuatara, and certain lizards and snakes.

ACRODONT DENTITION Identical teeth positioned in rows around the jaws, unsocketed and attached directly to the bone

ACROMIAL Refers to the CLAVICLE in anurans.

ACROMIAL PROCESS A small projection directed anteriorly on the cranial border of the SCAPULA and which, in chelonians, becomes quite extensive and lengthened, forming an essential part of the extraordinary PECTORAL GIRDLE found in this group.

ACTIVITY AREA *See* HOME RANGE.

ACTIVITY RHYTHMS The intermittent and repetitive pattern in the locomotory activity of organisms. Usually linked to a day/night variation (CIRCADIAN RHYTHM), and occurring both as ENDOGENOUS and EXOGENOUS patterns, and also consisting of different single activities.

ACUMINATE Ending in a tapering, sharp point.

ACUTE Sharply angled; having a sharp end or point.

ad. Abbreviation for ADULT.

ADAPTATION In organisms, an innate or acquired alteration such as a physiological process, behaviour trait, or anatomical feature for example, that makes those organisms better suited to survive and reproduce in a given environment.

ADAPTIVE BREAKTHROUGH Any evolutionary change, resulting from the attainment of a characteristic ADAPTATION, that allows a TAXON or population to progress from one ADAPTIVE ZONE to another, for example from water to land.

ADAPTIVE RADIATION The progress of two or more related groups of animals down entirely different evolutionary directions. This type of divergent evolution occurred in the MESOZOIC reptiles.

ADAPTIVE ZONE A TAXON that is considered equally with its specific habitat or niche. The adaptive occupation that matches the taxon to its environment (thus the adaptive zone), may be broad (as with the Indonesian blue-tongued skink *Tiliqua gigas* which feeds on a variety of foods) or narrow (as with Australia's thorny devil *Moloch horridus* which eats only ants).

ADDLED A term used to describe eggs that fail to hatch.

ADETOGLOSSAL Free-tongued. Used mainly in reference to the tongues of salamanders of the genus *Plethodon*

ADHERENT Attached, joined, sticking, related, or connected to.

ADHESION SURFACE Part of a subdigital LAMELLA in lizards, positioned laterally and sustaining many hair-like structures forming a fibrous group at its extremity.

ADHESIVE APPARATUS *See* ADHESIVE ORGAN.

ADHESIVE ORGAN; ORAL SUCKER Referring to the folds at the hindmost border of the mouth in embryonic tadpoles of anurans. These folds can be horseshoe-shaped, V-shaped or crescentic, or divided into two bilateral parts, and are covered internally with intricate mucus-secreting cells. This mucus enables the tadpoles to adhere to submerged objects. The organ is lost before METAMORPHOSIS.

ADHESIVE PAD Particular tissue occurring on the toes of certain lizards that has developed in such a way as to enable the animals to hold on to smooth vertical surfaces.

ADIPOSE Pertaining to fat.

ADIPOSE TISSUE Any spongy fat-retaining tissue of the body.

ADJUVANT MIXTURE The addition of any substance which boosts the effects of the primary component.

ADNATE Referring to two different organs that are closely connected, e.g., the tongue and the floor of the mouth in certain amphibians, chelonians and crocodilians.

ADNEXUS (pl.) **ADNEXAE** An additional and supplementary part to a principal organ or structure, e.g., an ovary is an adnexus of the uterus, and the eyelids and NICTITATING MEMBRANE are adnexae of the eye.

ADOPT To make use of an UNAVAILABLE NAME as the VALID NAME of a taxon in a way which determines it as a new name with its own authorship and date.

ADPRESS To press close to or hold against. As a method of determining the relative limb length in urodeles, the fore- and hind limbs on one side are held flush with the side of the body and the length of the extension of the toes, or the distance separating them, is measured. In anurans the hind limb is brought forward in line with the straightened body and the distance between the nostril and the heel is measured.

ADPRESSED LIMBS *See* ADPRESS.

ADRENAL CORTICOID ACTIVITY The action of hormones produced by ADRENAL GLANDS that create many effects including dilation and constriction of blood vessels, emotional state corrections, and metabolism of various materials and systems.

ADRENAL GLAND An endocrine gland at the anterior end of each kidney.

ADRENALINE A hormone produced by the adrenal medulla in reaction to stress. It increases blood pressure and pulse and heart rate, and raises the levels of glucose and lipids in the blood. It is extracted from animals or synthesized and used for various medical purposes such as the treatment of ANAPHYLAXIS following the bite of a venomous snake.

ADULT In herpetology, an individual which has attained the point in its life from whence its general shape and appearance ceases to alter with age, as opposed to a JUVENILE or LARVA.

ADVERTISEMENT A form of demonstration in which an individual makes itself as noticeable as possible and used most frequently by males maintaining a territory in order to fend off rivals and catch the attention of females.

AEROBE *See* AEROBIC.

AEROBIC Having a dependency upon air or free oxygen, i.e, an aerobe, or aerobium

AEROMONAS INFECTION Probably the most significant bacterial disease to occur in reptiles and amphibians, particularly in those living an aquatic existence. In reptiles, *Aeromonas hydrophila* infects mainly the digestive system, but can also be responsible for abscesses and pneumonia. In anurans, the same bacterium is responsible for the usually fatal disease known as RED-LEG.

AESTIVAL Pertaining to the early summer. *Compare* AUTUMNAL; HIBERNAL, PREVERNAL; SEROTINAL; VERNAL.

AESTIVATION A condition of

torpor during extended periods of high temperatures or drought; a state of inactivity during which the metabolic processes are greatly reduced.

AFIBRINOGENEMIA A disorder of the blood characterized by the absence of fibrinogen in the plasma resulting in incoagulability.

AFROTROPICAL *See* ZOOGEOGRAPHICAL REGION.

AGAMA Any lizard of the family AGAMIDAE.

AGAMIC Reproducing without fertilization; asexual.

AGAMIDAE Agamid lizards. Family of the SQUAMATA, suborder SAURIA, inhabiting much of the Old World (central, south and South-East Asia, and Africa except Madagascar) and Australia. One species occurs in Europe. Over 300 species in some 30 genera.

AGAMOGENESIS Asexual reproduction such as PARTHENOGENESIS.

AGGREGATE; AGGREGATION An assemblage of from several to many members of the same species amassed in the same place. In reptiles and amphibians this occurrence is generally connected with hibernation, e.g., wintering groups of rattlesnakes (*Crotalus*), or reproduction, e.g., courting groups of frogs (*Rana*).

AGGRESSION Any action of an animal that results in intimidation on, or injury to another animal, but which is not associated with true predation (e.g. the hunting and killing of that animal for food).

AGLOSSAL Having no tongue. Frogs of the family PIPIDAE are aglossal.

AGLYPH A term applied to any one of a large group of non-venomous snakes that have teeth without grooves, in contradistinction to the opisthoglyphous snakes. Teeth of aglyphous (or aglyphic) species are solid and unable to transfer venom.

AGLYPHIC *See* AGLYPH

AGLYPHOUS *See* AGLYPH.

AGONISTIC Pertaining to defensive or aggressive actions between two rival individuals of the same species.

AHRIMAN In Persian mythology, a snake-like being devoted to evil and whose kiss was likened to the venom of a snake. Ahriman was said to be responsible for introducing disease, uncleanness and cannibalism to the world.

AIR SAC The hindmost section of the lung of a snake. Besides increasing the efficiency of gaseous exchange within the lung, it could also act as an evaporative cooling surface or assist buoyancy in aquatic species.

AKINETIC Used in reference to a skull in which all the bones are attached inflexibly to each other thereby preventing movement between them.

ALA (pl.) **ALAE** A wing or flat wing-like structure or process, such as a part of some cartilages and bones.

ALAR Pertaining to, resembling, or possessing wings or ALAE.

ALAR PLATE In the snake genus *Eryx*, an accessory, flattened wing-like process extending up and back from the middle of the zygapophysial ridge on posterior caudal vertebrae. *See* ZYGAPOPHYSIS.

ALARM CALL *See* MALE RELEASE CALL.

ALARM REACTION Any one of a number of responses to sudden exposure to various situations in which the animal concerned feels threatened or in danger. In amphibians, for example, this reaction can take the form of mucus secretion, colour change or urination.

ALBINISM A hereditary condition reflected in an animal's hair, skin and eyes, resulting from the failure or inability to form melanin due to a shortage in the enzyme tyrosinase.

ALBINO An animal in which all dark pigment has failed to develop so that it appears white or pinkish with red or pinkish eyes. Albinism occurs as a genetic mutation in most groups of reptiles and amphibians.

ALBUMEN; ALBUMIN 1. Any of a group of uncomplicated water-soluble proteins that are coagulated by heat and which occur in blood plasma, egg white etc. **2.** The white of eggs in certain reptiles; the watery protein solution between the shell and the yolk.

ALBUMINURIA The presence of ALBUMEN in the urine; a characteristic symptom of envenomation by Russell's viper *Daboia russelli*.

ALETHINOPHIDIA One of two major lineages to which snakes belong. The alethinophidians or, 'true' snakes, consist of some 10

families totalling approximately 2400 species with a wide variety of habits and occupying a wide variety of habitats. The infraorder forms some nine tenths of all snake species and represents the snakes that once were subterranean burrowers and which re-emerged and readapted to life upon the surface. Continued diversification eventually gave rise to the HENOPHIDIA and later the CAENOPHIDIA. *See also* SCOLECOPHIDIA.

ALETHINOPHIDIAN *See* ALETHINOPHIDIA.

ALIEN *See* EXOTIC; INTRODUCTION.

ALLANTO-CHORION An embryonic membrane.

ALLANTOIC BLADDER A sac formed from the posterior region of the alimentary canal functioning, in amphibians, as a urinary bladder and, in embryonic reptiles, a receptacle for metabolic wastes.

ALLANTOIC GILL The respiratory construction occurring in the embryos of entirely terrestrial plethodontid salamanders which gives them a surface for gaseous exchange within the egg capsule. The gills ATROPHY when the larvae hatch.

ALLANTOIS One of the three embryonic membranes of reptiles and higher vertebrates. It arises from the hind gut in the developing reptile embryo and is used partly as an organ of nutrition and respiration and partly as a receptacle for storing waste.

ALLERGY A condition in which a

substance which is normally undisruptive causes acute sensitivity.

ALLIGATOR Crocodilian of the New World (*Alligator mississippiensis*) and China (*Alligator sinensis*), having a broad, flat snout and certain teeth in the lower jaw that fit into pits in the upper jaw, not notches as in crocodiles.

ALLIGATORIDAE Family of the CROCODILIA, the alligators and caimans, inhabiting the Americas and eastern Asia. Seven species in four genera.

ALLOCHTHONOUS Term given to an animal that has migrated into, or been transported and released into, a region that is not its usual habitat. The Neotropical toad, *Bufo marinus*, is allochthonous, having been introduced into a number of countries throughout the world.

ALLOPATRIC Applied to two or more animal populations that could interbreed but do not as they occupy mutually exclusive, although usually adjacent, geographical regions. *Compare* PARAPATRIC; SYMPATRIC.

ALTERNATE REPLACEMENT *See* INTERCALARY REPLACEMENT.

ALVEOLAR 1. Referring to the margin of the jaw where teeth, if present, are situated. **2.** Consisting of, or containing, many small hollows, sockets or pits.

ALVEOLAR RIDGE In chelonians, the ridge on the crushing surface of the upper jaw.

AMBIENT TEMPERATURE The temperature of the air or immediate surroundings, or of the enclosed and artificial environment of a captive animal.

AMELANISIM Lacking the pigment MELANIN which produces mainly black and brown colours. Reptiles that are amelanistic are consequently paler than normal, although they may have a certain amount of coloration from other pigments, particularly red or pink.

AMELANISTIC *See* AMELANISM.

AMINO ACID Any one of a group of nitrogen-containing organic acids which together form the building blocks of all proteins.

AMINOGLYCOSIDE In reference to the treatment of reptile and amphibian diseases, a group of antibiotics used for GRAM-NEGATIVE BACTERIAL infections. Gentamicin, for example, is an aminoglycoside.

AMMONIO-TELISM The physiological condition known to occur in several totally aquatic amphibian species, and the alligator, in which the main product of kidney excretion is ammonia. The majority of aquatic reptiles, such as turtles, excrete equal quantities of both ammonia and UREA, as do several species of amphibians.

AMNION One of three membranes formed in which the embryo develops within the egg. *See* ALLANTOIS; CHORION.

AMNIOTE Any vertebrate whose embryos are completely enveloped in a fluid filled sac, the AMNION. As this sac evolved it supplied the essential liquid environment for the embryo, enabling animals to breed in places other than water.

AMNIOTIC CAVITY A closed, fluid-filled pouch that protects the reptile embryo (and embryos of birds and mammals also).

AMOEBIASIS; AMOEBIC DYSENTERY; ENTAMOEBIASIS A PROTOZOONOSIS that can infect reptiles, particularly snakes and lizards but, to a lesser degree, chelonians also, causing slimy and watery faeces, which sometimes contain blood, refusal to feed, regurgitation and the urge to drink frequently. In humans, the most common form of amoebiasis is amoebic dysentery which can vary in its severity from mild diarrhoea to severe or even fatal dysentery.

AMOEBIC DYSENTERY *See* AMOEBIASIS.

AMPHIBIAN Any member of the vertebrate class Amphibia, containing the newts and salamanders (URODELA), frogs and toads (ANURA), caecilians (APODA), and sirens (TRACHYSTOMATA). Amphibians have four PENTADACTYL limbs, a moist and scale-less skin, and a pelvic girdle joined to the vertebral column. They are POIKILOTHERMS and differ from reptiles by producing eggs that have no protective shell or embryonic membrane and which, with the possible exception of a very few species, are fertilized externally.

ADULT amphibians have lungs and live mostly on land but their skin, which is also used in respiration, is thin and moist and, as body fluids are easily lost, they are usually dependent upon damp habitats. The class Amphibia falls intermediate between fishes and reptiles.

AMPHIBIAN MALARIA *See* FROG MALARIA.

AMPHIBIAN MUCORMYCOSIS *See* MUCORMYCOSIS.

AMPHIBIOTIC 1. Descriptive of an organism that can be either symbiotic with or parasitic on a specific host organism. 2. Inhabiting water (as a larva) and later land (as an adult).

AMPHIBIOUS Able to occupy both land and water. Usually used to describe the lives of most amphibians but is equally appropriate to semi aquatic chelonians, crocodilians and certain snakes and lizards.

AMPHI-DICHOTOMY In snakes, an ABERRANT condition in which the front and rear parts of the body are duplicated about a single, central trunk.

AMPHIGONIA RETARDA Delayed fertilization. In many reptiles and some amphibians sperm can remain viable for long periods, even several years, within the female before fertilizing the eggs.

AMPHIGYRINID The name given to any anuran tadpole that has two lateral SPIRACLES rather than one, as in the Aglossa for example.

AMPHISBAENA 1. Astounding poisonous serpent of classical mythology with a head at both ends and able to move forwards or backwards. **2.** GENUS of wormlike lizards.

AMPHISBAENIAN; AMPHISBAENID A worm lizard of the family AMPHISBAENIDAE.

AMPHISBAENIDAE Family of the SQUAMATA, suborder Amphisbaenia, inhabiting southern Europe, South America, and Africa. Around 120 species.

AMPHIUMA An aquatic, eel-like, nocturnal salamander of the genus *Amphiuma*, family AMPHIUMIDAE, with three species all restricted to the southeastern United States.

AMPHIUMIDAE Eel newts. Family of the URODELA, inhabiting southern and south-eastern North America. Three species in one genus. Amphiumas are eel-like in appearance and possess four ineffective tiny limbs each with one to three digits. Upon hatching the larvae have external gills and do not metamorphose completely, the adults losing the gills but retaining a pair of gill slits.

AMPLECANT Used to describe amphibians that are engaged AMPLEXUS.

AMPLEXATION *See* AMPLEXUS.

AMPLEXUS The sexual embrace of certain species of male amphibians upon the females. There are two basic types of amplexation: AXILLARY (pectoral) and INGUINAL (pelvic), and a number of variations including cephalic, lumbar, lumbo-pubic, neck, and supra-axillary.

AMPULLA URETERIS The swollen and expanded area in the URETER of snakes that acts as a vesicle for the storage of sperm during the mating season.

AMYOTROPHIN A powerful TOXIN, found in the venom of cobras (ELAPIDAE), that destroys muscles and nerve cells.

ANACHORESIS The habit of dwelling in crevices or holes in order to escape the attentions of predators.

ANAEMIA A condition common in debilitated reptiles. Symptoms include pale mucous membranes, listlessness and general weakness. An accurate veterinary diagnosis is vital. Parasite infestations are one likely factor as are acute RENAL or HEPATIC problems.

ANAGOTOXIC Has been used to describe anything that has the ability to neutralize the effects of venom such as, for example, the sulphurous waters of Sao Pedro in Brazil, and the root of the acafrao plant from the rainforests of Minas Gerals state in eastern Brazil which contains 'ar-turmerone' which has been shown to neutralize the venom of both the tropical rattlesnake and a dangerous local pit viper.

ANAL In reference to the anus; area of the body at and immediately surrounding the anus. 1. The posterior most LAMINA, usually paired, on the PLASTRON of many chelonians. 2. In snakes, the single, or divided, terminal plate of the VERTEBRAL series. Usually larger than the VENTRAL scales. Also known as pre-anal or post-

12

abdominal.

ANAL CLAW *See* ANAL SPUR.

ANAL FLAP *See* ANAL FOLD.

ANAL FOLD; ANAL FLAP In certain species of frog, a single fold of skin situated laterally and dorsally to the anus. In some species it is paired and situated laterally to the anus.

ANALGESIA; ANALGIA The inability to feel pain. A pain-killing drug is an analgesic.

ANAL GLAND A usually paired, bag-like feature located within the base of the tail in both sexes of snakes, though more prominent in females, and opening into the VENT, or anus. Used mainly in defence, it produces an evil-smelling fluid which can sometimes be sprayed a considerable distance.

ANAL PLATE; CLOACAL PLATE The plate, or SCUTE, which covers the VENT. In most lizards and snakes it is usually quite distinct being larger than the other VENTRAL scales.

ANAL RIDGE Referring to the KEEL on the scales directly above the anus of sexually mature male snakes of certain species which otherwise have smooth scales on the rest of their bodies. The anal ridge is absent in some adult males and may show in females so is not therefore a reliable guide to sex determination.

ANAL SPUR; ANAL CLAW; PELVIC SPUR A remnant of the hind limbs, usually visible on either side of the vent in the boas and pythons (BOIDAE) and Oriental pipe snake (*Cylindrophis*). In most

males the spurs are generally quite large and are often used as organs of stimulation during courtship. In females, they are small or, in some species, absent altogether.

ANAL WART *See* WART.

ANAMNIOTE Any vertebrate that is without an AMNION. Amphibians are anamniotes having embryos and larvae that must develop in water.

ANAPHYLACTIC SHOCK *See* ANAPHYLAXIS.

ANAPHYLAXIS In relation to the effects of certain venomous SNAKE BITES, a severe allergic reaction which may produce respiratory, circulatory, and neurological symptoms. Anaphylactic shock is frequently fatal if not treated.

ANAPSID Describing a skull that has no openings in the temporal region. Found in extinct reptiles (Cotylosauria) and in the present-day TESTUDINIDAE.

ANAPSIDA A subclass of the REPTILIA categorized by a skull lacking gaps in the temple area behind the eyes. Found in two or three extinct groups but represented today only by the CHELONIA.

ANATOXIN *See* ANAVENOM.

ANAVENIN *See* ANAVENOM.

ANAVENOM; ANATOXIN; ANAVENIN Snake venom which, after having been processed in order to remove its toxic properties, is injected into the blood of a living animal where antibodies are then formed, thus conferring an immunity to the anavenom. Blood serum from the immunized animal can then be used in the treatment of

SNAKEBITE.

ANCIPITAL Sometimes seen in early literature and used to describe the laterally compressed tails of certain lizards, newts and salamanders, which are double-edged.

ANELYTROPSIDAE Snake lizards. Family of the SQUAMATA inhabiting Mexico. MONOTYPIC.

ANERYTHRISM Lacking all red pigment. An anerythristic specimen is one that, through genetic mutation, lacks all red pigment so that any red markings normally present in that particular species are absent.

ANERYTHRISTIC *See* ANERYTHRISM.

ANGEL, Fernand, (1881-1950), French herpetologist who worked on *Mission scientifique au Mexique et dans l'Amerique Centrale: Etudes sur les reptiles et les batraciens* (1870-1909),*Vie et moeurs des amphibiens* (1947), and *Vie et moeurs des serpents* (1950), amongst others.

ANGLE OF THE JAW The angle created by the joint of the compound bone of the lower jaw with the quadrate bone of the skull.

ANGLE OF THE MOUTH; ANGULUS ORIS The angle created by the connective muscle tissue at the point where the upper and lower jaws diverge. In snakes and lizards the angle of the mouth is covered by the hindmost lower and upper LABIAL scales.

ANGUIDAE Alligator lizards, glass lizards and related species. Family of the SQUAMATA, suborder SAURIA, inhabiting Europe (including Britain), North and South America, West Indies, south and South-east Asia and the Middle East. 75 species in eight genera.

ANGUIFORM; ANGUINIFORM Snake-like; snake-shaped.

ANGULAR GLAND In some chelonians and in crocodiles, the MUSK GLAND situated on the throat on the inside of the left and right halves of the lower jaw.

ANGULATE Having angles or an angular shape. The head of the Pacific boa (*Candoia*) can be described as angulate.

ANGULUS ORIS *See* ANGLE OF THE MOUTH

ANILIIDAE Pipe snakes. Family of the SQUAMATA, inhabiting northern South America (one species), and India. Ten species in three genera.

ANISODONT Having teeth of unequal lengths.

ANKYLOSE; ANCHYLOSE To fuse together into one; to unite two bones into one. Ankylosis is the type of rigid bony union of tooth and jaw found in most toothed reptiles, except crocodiles. The teeth of most lizards and all snakes are ankylosed to the inner margin of the jawbone or in shallow notches (PLEURO-DONT). In certain lizard families the teeth are ankylosed to the top of a bony ridge within the jawbone (ACRODONT).

ANKYLOSIS *See* ANKYLOSE

ANNIELLIDAE Legless lizards. Family of the SQUAMATA, suborder SAURIA, inhabiting

14

California and Baja California, Mexico. Two species in one genus.

ANNUAL CYCLE In reference to the lives of reptiles and amphibians, the succession of biological incidents that take place each and every year.

ANNUAL RING *See* GROWTH RING.

ANNULAR HORN A horn made up of rigid tapering rings or ANNULI.

ANNULUS (pl.) **ANNULI** Any ring-shaped space, part or structure, such as any of the external body segments of the AMPHISBAENIDAE family of lizards.

ANOLE Any lizard of the largest existing lizard genus, *Anolis*, family POLYCHROTIDAE, inhabiting the south-eastern, Central, and South America and many islands of the Caribbean.

ANOMALEPIDAE Blind snakes. Family of the SQUAMATA, suborder SERPENTES, infraorder Scolecophidia, inhabiting tropical South America. Some 20 species in four genera.

ANONYMOUS From or by a person, author (s), etc., whose name is unknown or withheld.

ANTARCTIC The faunal region which, besides the Antarctic continent itself, includes the Falkland Islands, Kerguelen Islands and the southernmost tip of South America. There are no RECENT amphibians or reptiles in Antarctica itself and only a handful (some larva lizards of the Tropidurinae) have spread into Tierra del Fuego from the Neotropics.

ANTE-ANAL LAMELLA *See* PREANAL.

ANTEBRACHIAL 1. Referring to any one of the scales situated on the frontal margin of the upper forelimb (the brachium). **2.** Any of the scales situated on the lower forelimb (the ANTEBRACHIUM).

ANTEBRACHIUM; ANTIBRACHIUM The section of the forelimb that contains the ULNA and RADIUS bones; the lower forelimb.

ANTEOCULAR *See* PREOCULAR.

ANTEORBITAL *See* PREOCULAR.**ANTERIOR** Located at, or towards, the front; of, or near, the head. *Compare* POSTERIOR.

ANTERIOR FRONTAL *See* INTERNASAL.

ANTERIOR LOBE The section lying anteriorly to the AXILLARY NOTCH of a chelonian PLASTRON.

ANTERIOR ORBITAL In snakes, the PREOCULAR.

ANTERIOR PARIETAL Used in reference to the first large scale behind the OCULAR in snakes of the ANOMALEPIDAE.

ANTIBIOTIC Any one of various chemical substances, such as penicillin or tetracycline for example, produced by various micro-organisms, especially fungi, or manufactured synthetically and capable of inhibiting or destroying the growth of other micro-organisms, especially bacteria.

ANTIBOTHROPHIC An ANTIVENIN effective against venoms that are characteristic of the

NEOTROPICAL pit vipers other than rattlesnakes; e.g., *Bothrops* etc.

ANTIBRACHIUM *See* ANTEBRACHIUM.

ANTICOAGULANT Any agent which prevents the blood from clotting and promotes bleeding (compare COAGULANT). Certain snake venoms such as, for example, that found in the king brownsnake *Pseudechis australis*, possess an anticoagulant property.

ANTIFIBRIN An ANTICOAGULANT in the venoms of certain snakes that obstructs the change of FIBRINOGEN into FIBRIN in the blood clotting creation development.

ANTIHISTIMINE Any drug that counteracts the effects of HISTAMINE, used principally in the treatment of allergies and allergic reactions

ANTILLEAN Relating to the Greater (Jamaica, Cuba, Puerto Rico and Hispaniola) and Lesser (Virgin Islands to Aruba) Antilles.

ANTIPALMAR Opposite the palm; relating to, or situated on, the back of the forefoot.

ANTIPLANTAR Opposite the sole; relating to, or situated on, the back of the hind foot.

ANTISERUM; ANTIVENENE; ANTIVENIN; ANTIVENOM; SERUM; SNAKE ANTITOXIN; SNAKEBITE SERUM The serum (the fluid part of the blood that remains after clotting) of an animal that has been immunized against a foreign substance and which acts together with the substance against which it has been produced. Serum is processed from the blood of an animal that has received gradually increased amounts of VENOM or ANAVENOM, or a mixture of the two and, as a result, has become immunized against the venom of a specific snake species. Antiserum is produced in horses usually but also in rabbits and sheep and used in the treatment of ENVENOMATION in other animals including man.

ANTISPASMODIC A substance administered to prevent or arrest muscle spasms.

ANTIVENENE; ANTIVENIN *See* ANTISERUM.

ANTIVENOM *See* ANTISERUM.

ANURA; SALIENTIA; TAIL-LESS AMPHIBIANS The order of amphibians, that contains the frogs and toads. There are over 2600 species occurring throughout the temperate and tropical regions of the world. Most frogs (e.g., *Rana*) live in damp places or are aquatic; many are ARBOREAL. Toads (e.g., *Bufo*), which often have a dry warty skin, are better adapted to drier habitats. Adult anurans have no tail, a short backbone with the head joined directly to the trunk, relatively long, powerful and webbed hind limbs and are agile swimmers, climbers and jumpers. There are exterior ear membranes posterior to the usually large eyes which are situated high on the head. SEXUAL DIMORPHISM is common, the males frequently being smaller and possessing VOCAL POUCHES. Anuran fertilization is external and eggs

16

(SPAWN) are covered with a protective jelly and, in most species, deposited in water in strings or clumps. The aquatic larva, or TADPOLE, experiences a normally quick and complete METAMORPHOSIS in which the GILLS are replaced by lungs and the tail absorbed. *See* AMPHIBIAN.

ANURAN *See* ANURA.

AODAISHO The Japanese ratsnake *Elaphe climacophora*, of the family COLUBRIDAE, an albino colour morph of which is often associated with the goddess of love and fertility.

APICAL An alternative term for the ROSTRAL when referring to snakes of the genus *Vipera*.

APICAL BRISTLE *See* TACTILE BRISTLE.

APICAL PIT; SCALE FOSSA; SCALE PIT A sense organ set in a small depression on the posterior tip of the DORSAL scales of certain kinds of reptiles. Sometimes called apical pore.

APICAL PORE *See* APICAL PIT.

APLASIA The failure of an organ to develop.

APNEA A pause in breathing, usually temporary; frequently a result of stimuli reduction to the respiratory centre.

APNEUMONIC Lungless. Used in reference to salamanders of the genus *Plethodon*.

APODA The limbless amphibians - CAECILIIDAE, or Gymnophiona - an extant order of tropical and sub-tropical worm-like animals of the class Amphibia. Adapted to a subterranean life, caecilians have rudimentary eyes below which are sensory feelers and a vestigial hearing system. They lack limbs and limb girdles and their tail is small or absent altogether. They are the only living amphibians in which certain species have retained scales, although these are also VESTIGIAL and concealed in skin folds. Males have cloacal copulatory organs and, unlike most other amphibians, fertilization is internal.

APODAL 1. Without limbs. Snakes, amphisbaenids and caecilians are apodal. **2.** Without feet. Describing lizards that have shortened stub-like limbs, as have certain members of the SCINCIDAE and TEIIDAE families.

APOSEMATIC COLORATION Describing the coloration of certain poisonous or distasteful animals (aposematism) which gives warning to possible enemies or predators. Examples of aposematic coloration are seen in the vivid blacks, yellows and reds of dart-poison frogs (DENDROBATIDAE) and fire bellied toads (DISCOGLOSSIDAE).

APOSEMATISM *See* APOSEMATIC COLORATION.

APPENDAGE Any projection from the body, or trunk, of an animal, such as a crest, horn or limb.

APRON A pronounced skin fold located on the lower throat in certain species of *Bufo*, which rests upon the folds of the VOCAL SAC when it is deflated.

AQUA-TERRARIUM Type of housing used for amphibious or aquatic amphibians and reptiles. An aqua-terrarium, in contrast to a straightforward aquarium, always has a land area on which the occupants can find shelter or bask etc.

AQUATIC Dwelling, developing, or occurring in water.

ARACHNID Any one of a class of invertebrates which includes the scorpions, spiders, ticks and mites. Both ticks and mites (ACARIDS) are of significance in herpetology as they are common parasites of reptiles. *See* ACARIASIS.

ARAZPA Acronym denoting the Australian Regional Association of Zoological Parks and Aquaria.

ARBOREAL Dwelling in, or having the ability to move in, trees and bushes. The term is also used by some authors to describe species that inhabit scrub, bush or rainforest etc.

ARCHIPELAGO A group of islands.

ARCHOSAURIA A subclass of the more advanced DIAPSID reptiles that includes the orders Crocodilia, Ornithischia, Saurischia, Pterosauria, and Thecodontia.

ARCTOGAEA A major region comprising the four northern zoogeographical regions of the Earth: the ETHIOPIAN, ORIENTAL, PALAEARCTIC and NEARCTIC regions. The Palaearctic and Nearctic regions are occasionally called the HOLARCTIC region as the fauna occurring in both is similar.

ARCUATE Curved or arched.

ARENICOLOUS Living in sand or inhabiting sandy places. Used to describe such desert species as the

sandfish *Scincopus fasciatus* and horned vipers (*Cerastes*).

AREOLA 1. A region of small elevated and rounded lumps on the skin, frequently seen on the belly in certain species of frogs, when it sometimes referred to as 'granular'. **2.** The central region of a LAMINA on the shell of a chelonian. It can be either raised or hollow and is the granular juvenile scute from which new growth extends outwards.

ARGASIDAE *See* TICK.

ARID Having little or no rain and referring to the typical climate of a particular region, which decides the state of the soil and flora and fauna within that region.

ARID TROPICAL SCRUB *See* TROPICAL ARID FOREST.

ARKS Acronym denoting the Animal Records Keeping System. ARKS is a software package for keeping accurate animal inventory records at individual facilities and institutions. It provides a framework for keeping data that is compatible with the ISIS database. The programme also generates inventory reports on individual animals, species, and entire zoological collections.

ARMY A collective noun for a group of frogs.

AROUSAL An increase in the awareness of an animal to sensory stimuli.

ARRENOIDISM Used to describe the condition in which females possess characteristics which are normally typical attributes of males. In *Bothrops insularis*, for example, a proportion of reproducing females have been found to have a HEMIPENIS.

ARRIBADA A concurrent, mass emergence of marine turtles on to a beach to deposit their eggs. Arribadas are influenced by the phases of the moon.

ARTHROPOD *See* ARTHROPODA.

ARTHROPODA An extremely varied PHYLUM of paired, jointed-limbed animals, possessing a segmented body and an exoskeleton of chitin. Included amongst it are the crustacea, insects, arachnids, and related forms, and together they contain over three quarters of all presently known animal species.

ARTIFICIAL CLASSIFICATION In zoology, CLASSIFICATION established by considering the more convenient or prominent indicative features of an organism rather than to those signifying association.

ARTIFICIAL INSEMINATION The artificial introduction of semen into the reproductive tract of a female by means other than sexual union. This method has been used successfully in the captive breeding of certain, so-called 'difficult' species where the animals have refused, for one reason or another, to reproduce naturally.

ARTIFICIAL RESPIRATION Any of various life-support methods of restarting breathing after it has stopped, to ensure adequate blood oxygenation, by manual rhythmic pressure on the chest, mouth-to-mouth breathing, or mechanical ventilation.

ARTIFICIAL SELECTION The selection by humans of specific animals (or plants) from which to breed the next generation, because they have the most obvious development of the necessary traits. Usually, the process is repeated in consecutive generations until those traits are fixed in the descendent offspring.

ASCAPHIDAE Tailed frogs. Family of the ANURA, inhabiting Pacific North America. MONOTYPIC.

ASCARID Any one of a group of nematode worms of the family ASCARIDOIDEA.

ASCARIDOIDEA Internal parasitic roundworms of the class NEMATODA. Although they are common in many lizards they are mainly found in the BOIDAE (the boas and pythons).

ASCI Acronym denoting an Area of Special Conservation Interest.

ASCORBIC ACID Vitamin C present especially in citrus fruits, tomatoes and green vegetables and required in the diet of certain captive reptiles.

ASPECT The compass direction in which an area of land faces. Reference to the aspect of an area is a significant element in the description of its habitat, particularly in respect of egg-laying sites and HIBERNATION quarters.

ASPERITY A roughened surface or outgrowth, sometimes used as an alternative to NUPTIAL PAD in male anurans.

ASPIRATE The inhaling of fluids into the lungs.

ASPIRATOR A suction device for removing fluids from a body cavity. Such a piece of equipment is often carried by field herpetologists for applying to a venomous snake bite in the accidental event of them being bitten.

ASRA Acronym denoting the Association for the Study of Reptilia and Amphibia (United Kingdom).

ASSERTION DISPLAY A somewhat weak and casual display in certain lizard species which is not always aimed at another of its own species and, indeed, frequently occurs when no other individuals are present.

ASYMPTOMATIC Showing no symptoms.

ATAXIA Loss of muscle coordination.

ATELOPODIDAE Family of the ANURA, suborder Procoela, inhabiting Central and South America. Over 30 species in two genera.

ATLASSING The collection of information on the distribution of reptiles and/or amphibians in a particular region for use in the compilation of an atlas.

ATRACTASPIDAE The mole vipers or side-stabbing snakes. Family of the SQUAMATA, suborder SERPENTES, genus *Atractaspis,* formerly placed in either the Viperidae or more recently the Colubridae.

ATRIO-VENTRICULAR Collection of adapted cardiac muscle fibres forming part of the heart's stimulus conductance system.

ATROPHY The reduction in size of an organ or part, as a result of disease, poor nutrition or lack of use. The legs and feet of slow worms (ANGUIDAE) have, for example, through the course of evolution, become atrophied as the lives of these lizards have become more subterranean.

ATTENUATE Slender and tapering to a sharp point.

AURICULAR 1. Of, or relating to, the auricles of the heart.
2. Of, or relating to, the ear and generally relating to external features. In many lizards the auriculars are modified scales that protrude over the front of the ear opening.

AURICULAR FOLD The fold of skin, which often has spines or enlarged knoblike projections, that lies behind or over the ears of certain lizards.

AURICULAR LOBULES The small projecting scales on the front edge of the ear opening in some skinks. *See* AURICULAR.

AUSTRALASIAN REGION One of the six zoogeographical regions of the world. Subdivided into three subregions, Australasia consists of the New Zealand subregion; Papua-Polynesian subregion (New Guinea and other South Sea Islands, of which Fiji and the Solomons are herpetologically important); and continental Australasian subregion (including Tasmania).

AUTHOR In herpetology, the person to whom a scientific name or work is accredited.

AUTHORITY In herpetology, **1.** The name of the AUTHOR of a taxonomic name, quoted after the name, e.g., the striped swamp snake *Limnophis bicolor bangweolicus* Mertens. **2.** Legal consent granted to an individual to undertake a specified action, e.g., collect, photograph, keep, or study protected wild reptiles and amphibians. **3.** A public board or corporation implementing governmental power either locally or nationally.

AUTHORSHIP The AUTHOR of a taxonomic name is the person who is solely responsible for both it and the specifications under which it is made available.

AUTOCHTHONOUS Opposite of ALLOCHTHONOUS. The existence of an animal (or plant) species within its HOME RANGE. The American bullfrog *Rana catesbeiana* is autochthonous in its natural range of eastern North America, but in the southern European Alps and Cuba, it is allochthonous.

AUTOHAEMORRHAGE Self-induced bleeding brought about by the intentional bursting of blood vessels and the rupture of outer surfaces. When boid snakes of the genus *Tropidophis* are threatened or disturbed, blood is oozed from membranes of the nostrils and mouth. Lizards of the genus *Phrynosoma* can eject blood from vessels in the corners of their eyes if they too are provoked or disturbed.

AUTOIMMUNITY The apparent capability of a snake to endure the effects of its own venom and that of others of its own or related species.

AUTONOMIC Relating to any part of the nervous system concerned with controlling instinctive body functions e.g., glands, smooth muscle, and the heart.

AUTONOMOUS Self-regulating. Used, in herpetology, to describe reptile or amphibian populations that maintain their numbers independently of outside influence.

AUTOPHARMACOLOGICAL SUBSTANCE Any one of a group of chemicals formed and released by body cells in reaction to a stimulus (e.g., a VENOM). Their production may result in adverse consequences, e.g., alterations in heart rate, shortness of breath, and shock.

AUTOTOMY Self division or fracture; the ability to cast off, spontaneously or by reflex, a part of the body and, in herpetology, usually referring to breakage and loss of the tail. Many lizards and certain urodeles can autotomize their tail if it is seized. In most cases a new tail is grown later, though never quite as perfectly as the original. In the Puerto Rican gecko *Sphaerodactylus roosevelti*, the process has evolved further in that the lizard's fragile skin has actually been designed to break leaving predators with a mouthful of skin rather than the gecko.

AUTOTOMY PLANE *See* BREAKAGE PLANE.

AUTOTOXIN Any substance produced within an organism that is TOXIC to that organism. Envenomation from the bite of certain snake species can frequently encourage the discharge of autotoxins in the victim that can produce even further damage.

AUTUMNAL Pertaining to the autumn. Compare AESTIVAL; HIBERNAL; PREVERNAL; SEROTINAL; VERNAL.

AVAILABLE NAME In nomenclature, any name that satisfies the ICZN's strict requirements, such as its formation, publication and date, and language etc. An available name is not automatically a VALID NAME whereas a valid name must always be an available one.

AVIDITY The swiftness with which the symptoms of a VENOM are eased by an ANTIVENIN.

AVITAMINOSIS *See* HYPOVITAMINOSIS.

AXIAL BIFURCATION The duplication of a part of the body along its length, sometimes occurring in chelonians and lizards but most commonly in snakes.

AXILLA Pertaining to the pit of the arm or forelimb.

AXILLARY Near or on the apex ; pit of the arm or forelimb.

AXILLARY EMBRACE *See* PECTORAL AMPLEXUS.

AXILLARY GLAND A glandular region, noticeable on the chest of pelobatid frogs of the genus *Megophrys*, at the insertion of the forelimb. In *M. hasselli* the gland is flat and rounded, and in *M. monticola* it is tubercle-like and cone-shaped.

AXILLARY LAMINA The cuticular SCUTE, or scutes, on the frontal border of the BRIDGE in chelonians

AXILLARY NOTCH The indentation in the chelonian shell through which the forelimb extends.

AXOLOTL The name, of Aztec origin, for any of the several larval salamanders of the family Ambystomidae which inhabit certain Mexican lakes. These larvae may live and breed as larvae (NEOTENY) but, under certain conditions, are capable of resorbing their fins and gills, and finally emerging from the water as air-breathing adult salamanders.

AZOTEMIA The presence of nitrogen, particularly urea, in increased quantities in the blood.

AZYGOUS; AZYGOS Occurring singly. Frequently used in reference to the enlarged scales, or plates, other than the normal ones present, on the heads of lizards and snakes, e.g., the third INTERNASAL in the genus *Heterodon*, the hognosed snakes, and the elongated scale located between the PREFRONTAL SCALES on the front of the upper head in certain sea snakes the presence or absence of which can be used to distinguish between *Laticauda laticaudata* and *L. colubrina*. (Termed 'azygous scale' by various authors).

AZYGOUS PLATE *See* AZYGOUS.

AZYGOUS SCALE *See* AZYGOUS.

B

BACK-FANGED *See* REAR-FANGED

BAIRD, Spencer Fullerton, (1823-1887), American naturalist responsible for establishing the herpetological collection at the national Smithsonian Institution in Washington DC. Co-wrote many publications with Charles F. Girard, including *Catalogue of North American reptiles* (1853).

BALANCER *See* STABILIZER.

BALE A collective noun for a group of turtles.

BALL *See* BALL POSITION.

BALL POSITION A defensive attitude assumed by certain snake species, namely the royal python, *Python regius*, often referred to as the 'ball python' and the Calabar ground python, or burrowing python, *Calabaria reinhardtii*, in which the body is coiled tightly into a ball. The tip of the tail is protruded from the coils and, in some cases, waved about in order to distract a potential enemy's attention from the vital and otherwise vulnerable head which is kept hidden in the centre of the ball. This defensive behaviour in the royal python *P. regius* usually disappears after a brief spell in captivity whereas it rarely does so in the burrowing python *C. reinhardtii*. Consequently, referring to the royal python as the ball python, as is frequently done in popular pet-care books and even some scientific papers, is not only confusing but also not entirely accurate.

BALL PYTHON *See* BALL POSITION.

BAND; CROSSBAND A particular pattern of skin colour that is darker or paler than the ground colour and which transverses the vertebral line. A band may or may not extend onto the ventrals, and may sometimes completely encircle the body. *See* RING.

BANDY BANDY The common name for any one of four or five mildly venomous, burrowing members of the family ELAPIDAE, subfamily Elapinae, genus *Vermicella*. They inhabit most regions of Australia except southern Western Australia, northern parts of the Northern Territory, and the far south-east. These small burrowing snakes are all similar in appearance, being strikingly banded with black and white rings which usually encircle the entire body. They feed almost exclusively on blind snakes (TYPHLOPIDAE) and because of their subterranean habits are rarely seen. The species include *Vermicella. annulata, Vermicella multifasciata, Vermicella intermedia,* and *Vermicella vermiformis*, all formerly just one species, *Vermicella annulata*.

BARBA AMARILLA; TERCIOPELO; TOMIGOFF; YELLOW BEARD *Bothrops asper*, one of the world's largest and most notorious species of PIT

VIPER. A member of the family VIPERIDAE, subfamily Crotalinae, inhabiting Central and South America, reaching up to 2400 mm in length and responsible for a high incidence of snake bite throughout its range. Often wrongly referred to as the FER DE LANCE.

BARBEL Any one of several small, fleshy downward-pointing projections on the throats and chins of certain chelonians, e.g., *Kinosternon*, functioning as a tactile organ. Such barbels also occur on the perimeter of the mouth in larvae of *Rhinophrynus* toads.

BARBOUR, Thomas, (1884-1946), American zoologist and director of Harvard University's Museum of Comparative Zoology. Initiated, with Leonard Stejneger, the *Checklist of North American amphibians and reptiles* (1917).

BARRED In reference to the patterns of snakes, possessing perpendicular markings of distinctly different colours on the side of the body.

BASIHYAL VALVE The lower of two muscular folds, or plicae, in the rear of the mouth of a crocodilian. Its function, in combination with the upper fold (VELUM PALATI), is to prevent water from entering the glottis when the mouth is opened below the water's surface.

BASILISK 1. A legendary monster, part snake and part cockerel; a cockatrice or fabulous reptile hatched by a serpent from a cockerel's egg, the glance and breath of which were supposed to be fatal. **2.** A Central and South American lizard of the family CORYTOPHANIDAE (formerly of the IGUANIDAE), genus *Basiliscus.*

BASKING A rather loose term used to describe the behaviour of a reptile or amphibian when exposing itself to the direct or diffused rays of the sun. THERMOREGULATION may or may not take place during periods of basking which are used for other functions, such as attaining a high body temperature, allowing greater mobility and aiding digestion.

BASKING LAMP A lamp or overhead heating element in an enclosure, or vivarium, that produces an area of concentrated heat required by a captive reptile for digestion.

BASKING RANGE The temperature range within which basking reptiles are mostly inactive whilst still remaining watchful. At the top of this range is the VOLUNTARY MAXIMUM temperature, or the thermal point at which the reptile becomes either uncomfortable and seeks shade or sufficiently warmed to carry out its normal activities. At the lower end of the range is the VOLUNTARY MINIMUM, the point where the reptile either seeks direct sunlight and more warmth or retires below ground in order to hibernate, for example.

BATES, Henry Walter, (1825-1892), The English naturalist who accompanied A. R. WALLACE on an exploration of the Amazon in 1848. His studies of the almost 15,000 species of insects he collected, of which some 8000 were new to science, led him to suggest the way in which the mimicry, subsequently named after him, can occur among unrelated species.

BATESIAN MIMICRY Named after Henry Walter BATES, the imitating of brightly coloured, or strikingly marked, unpalatable species by certain species which are palatable. By mimicking the pattern and colours of poisonous animals other not hurtful species may acquire a certain amount of protection against predators. In herpetology examples include the harmless red salamander *Pseudotriton ruber* which mimics the toxic immature larvae of the red-spotted newt *Notophthalmus viridescens*.

BATRACINE A highly toxic chemical occurring in the skin secretions of *Dendrobates* frogs.

BATRACHIAN An archaic term for any amphibian, especially a toad or frog; of the class Batrachia.

BATRACHOLOGY Although still used by some French authors, a rather outdated term used once when amphibians were studied separately from reptiles.

BEAD The slender, outwardly directed curve on the front side rim of a rattlesnake RATTLE that is shaped in such a way as to prevent splitting and acts also as a sound-making mechanism as it strikes the segment next to it when the tail is vibrated.

BEAK The horny covering (RHAMPHOTHECA) of the jaws or the horny beak-like mouthparts of tadpoles.

BEAK-HEAD *See* RHYNCHOCEPHALIA.

BED A collective noun for a group of snakes.

BELL An individual segment of a rattlesnake's RATTLE.

BELL, Thomas, (1792-1880**),** English zoologist. Published the definitive *Monograph of the Testudinata* (1836-1842) containing plates by Edward Lear and James de Carle Sowerby.

BELLAIRS, Angus d'Albini, (1918-1990**),** the father of herpetology in Britain. Taught embryology in the University of London at St Mary's Hospital Medical School and was Honorary Herpetologist to the Zoological Society of London. He produced numerous papers, particularly on the adaptations of vertebrates and their embryos to the problems of their environment. His books include *Reptiles* (1957), *The world of reptiles* (1966), with Richard Carrington, and the two-volume *The life of reptiles* (1969).

BELL GILL The greatly enlarged gill, which acts as a respiratory exchange surface, in tadpoles of certain hylid frogs, including *Gastrotheca* and *Cryptobatrachus*, which have a BROOD POUCH on their backs in which they carry their larvae.

BELLY The abdomen; the entire under surface, or VENTER.

BELLY INSIGNIA; BELLY PATCH The area of vivid and distinctive coloration in mostly mature male iguanids, situated on the flanks between the fore and hind limbs and revealed to rivals and females during territorial and courtship displays.

BELT TRANSECT A strip, normally 1 metre wide, that is marked out across a particular habitat and within which species can then be recorded in order to determine their distribution and/or population density. *See* TRANSECT. *Compare* LINE TRANSECT.

BENTHOS In marine and freshwater ECOSYSTEMS, the animals and plants living on the bottom of a sea or lake.

BERN CONVENTION The Convention on the Conservation of European Wildlife and Natural Habitats 1979. Founded by the Council of Europe to: 'conserve wild flora and fauna and their natural habitats; promote co-operation between states; monitor and control endangered and vulnerable species; and assist with the provision of assistance concerning legal and scientific matters'. The Convention lead to the establishment, in 1998, of the EMERALD NETWORK of Areas of Special Conservation Interest (ASCIs) throughout the territory of the parties to the Convention, which operates alongside the European Union's NATURA 2000 programme.

BETA-GLOBULIN Globulin proteins in blood plasma associated with certain antibodies. *See* GLOBULINS.

BEZOAR STONE A hard mass such as a hairball or stone occurring in the stomach. Formerly greatly prized as a medicine, bezoars were frequently used in the treatment of SNAKE BITE. Goats were once reared especially for these much sought-after stones.

BHS Acronym denoting the British Herpetological Society (United Kingdom).

BIBRON, Gabriel, (1806-1848), French herpetologist. Co-authored, with André Duméril and his son Auguste Duméril, the ten-volume *Erpetologie generale ou Histoire naturelle complete des reptiles.*

BIBRON'S ANTIDOTE A concoction once routinely supplied to the United States army and thought to be a remedy against the bites of rattlesnakes.

Bic Abbreviation for BICOLOUR.

BICARINATE *See* CARINATE.

BICEPHALOUS; DICEPHALOUS Having two heads. A phenomenon sometimes seen in reptiles, particularly snakes, where two fully-functional heads are present on the same body.

BICOLOUR Literally 'two colour'; usually relating to snake patterns characterized by an alternating RING or BAND of dark (generally black) and light (red, yellow or white), usually red as in the coral snake *Micrurus mipartitus*.

BICUSPID Possessing two CUSPS. Some pigmy chameleons (e.g., *Rhampholeon*) exhibit such features on their feet as too do the BEAKS of certain chelonians.

BIDDER'S ORGAN An undeveloped ovary occurring in the males of certain anurans.

BIENNIAL BREEDING CYCLE
The breeding cycle occurring in the females of certain snake species generally in temperate climates where conditions compel some of them to breed every two years.

BIFID Divided into two rounded projections, or forked, e.g. the tongue of all snakes and some lizards (VARANIDAE).

BILATERAL Possessing or involving two sides.

BILATERAL SYMMETRY The arrangement of the body parts of an animal in a way that when divided by a plane the two halves are mirror images of each other. Bilateral symmetry is linked with movement in which one end of the animal constantly goes in front.

BILIRUBIN The orange-red pigment in bile produced from red blood cell HAEMOGLOBIN.

BILOBED Divided into two lobes.

BINOCELLATE Having two eye-like spots, as on the rumps of certain leptodactylid frogs, e.g., *Telmatobius praebasalticus* and *Physalaemus nattereri*, which display them to frighten potential enemies.

BINOCULAR VISION A type of vision in which the eyes are directed forwards so that the image of an individual object can be focused on the retinas of both eyes simultaneously, thus granting perception of distance and depth. Snakes of the genus *Ahaetulla* have large, prominent eyes giving them acute three-dimensional vision with the ability to detect the slightest movement.

BINOMEN *See* BINOMIAL NOMENCLATURE.

BINOMIAL NAME *See* BINOMIAL NOMENCLATURE.

BINOMIAL NOMENCLATURE A system for naming animals (and plants) by means of two Latin names (binomen), the first, with an initial capital letter, indicating the GENUS and the second the SPECIES to which a particular organism belongs, as in *Pantherophis guttatus* (the corn snake). The strict measures and policy for naming newly discovered organisms are set by the INTERNATIONAL COMMISSION FOR ZOOLOGICAL NOMENCLATURE.

BINOMINAL Of, or denoting the BINOMIAL NOMENCLATURE.

BIOCELLATE Possessing markings resembling two eye-like spots or OCELLI.

BIOCLIMATOLOGY The study of how climatic conditions have an effect upon living organisms.

BIODIVERSITY In the main there are three kinds of biodiversity: genetic diversity (referring to the individuals of a species within a given population which breed and share their GENES with each other rather than with those in another population); habitat diversity (referring to the variety of environments in which life exists e.g., arid grassland, deciduous forest, coastal wetland, etc); and, most commonly, species diversity (referring to the many and varied different types of animals and plants locally, nationally, or worldwide, the term 'species' being a rank of CLASSIFICATION in the taxonomic ladder that includes the GENUS, the FAMILY, the ORDER, the PHYLUM, and the KINGDOM.

BIO-GEOGRAPHICAL BARRIER Any obstacle, natural or man made, preventing the migrating of species. Such barriers may be climatic (e.g., involving temperature, humidity, food availability etc.,) or physical (e.g., a lake, sea, or mountain range etc.).

BIOGEOGRAPHY The branch of biology dealing with the geographical distribution of animals and plants, subdivided into animal geography (ZOOGEOGRAPHY) and plant geography, or geobotany.

BIOLOGICAL CLOCK The ENDOGENOUS mechanism of an organism that produces regular periodic alterations in behaviour or body functions that are synchronized with environmental conditions only triggered by outside factors, as in HIBERNATION, when reptiles kept under artificial conditions either stop or reduce their food intake for a while during the months when they would be hibernating in the wild.

BIOLOGICAL SPECIES Said of animals that are distinct from others because, according to Ernst Mayr's principle, they cannot form fertile hybrids between each other naturally. This is particularly appropriate for those interested in population genetics, breeding behaviour, and the introduction of alien species. *See* EVOLUTIONARY SPECIES; TAXONOMIC SPECIES.

BIOMASS The total amount of all living organisms in a given population or inhabiting a given region, expressed in terms of living or dry weight per unit area.

BIOME A major ecological community extending over a large area and generally characterized by a dominant vegetation; SAVANNA and TUNDRA for example.

BIOSPECIES A group of interbreeding individuals that is remote reproductively from all other groups.

BIOSPHERE RESERVE Any one of the chain of conservation sites nominated by the United Nations Educational, Scientific, and Cultural Organization (UNESCO) in an endeavour to create an international system of protected areas encompassing examples of all the most important vegetation and physiographic types on Earth.

BIOSYNTHESIS The formation of complex compounds from simple substances by living organisms.

BIOTA The plant and animal life of a particular region or period.

BIOTELEMETRY; RADIOTELEMETRY A method for monitoring the movements of a reptile by means of a small, remote-sensing transmitter which is either attached externally to the body or implanted within the body cavity, enabling the accurate position of the reptile to be determined.

BIOTOPE The specific ecological area such as a river bank, gravel pit, sand dune etc., or region such as a desert, characterized by certain

28

conditions, and where characteristic animals or plants (biota) live.

BIOTYPE A group of animals within a species, occurring in the wild state, that resemble but differ physiologically from, other members of that species.

BIPEDAL Having two feet; a term used to describe lizards that can stand or run on their hind limbs with their forelimbs held off the ground, as in the frilled lizard *Chalmydosaurus kingii.*

BIPEDIDAE Two-legged worm lizards. Family of the SQUAMATA, suborder Amphisbaenia, inhabiting western Mexico. Three species in one genera.

BIRTH PLATE The LAMINA in chelonians that is present on the shell during the period of hatching and which, in certain species, is shed a short while after hatching, whilst in others it survives as an area encircled by growth rings.

BISERIAL Side by side.

BISHOP, Sherman C., (1887-1951), American herpetologist. Author of *Handbook of salamanders* (1943).

BLACK ALBINO An alternative term used occasionally for an ANERYTHRISTIC specimen.

BLACK SNAPPER *See* MASSASAUGA.

BLADE HORN In certain members of the CHAMAELEONIDAE, a fleshy horn made up entirely of scales and shaped like a blade.

BLEB A small blister containing blood or serum, frequently found on the ventral scales of captive snakes which are housed in conditions

which are too damp. Untreated it can soon develop into a BULLA which in turn can lead to secondary, and sometimes fatal, infections of the skin.

BLOOD FLUID VOLUME The level of blood usually circulating within the body.

BLOOD-RED An inherited species variety characterized by a relatively solid, dark, red-orange coloration and an abnormal ventral pattern.

BLOOD VOLUME DEFICIT The level of blood in effect lost to circulation within the body.

BLOOM *See* IRIDESCENCE.

BOA BUG *See* INCLUSION BODY DISEASE.

BOBBING The rapid, bouncy, up and down movements of many lizard species during territorial and courtship displays.

BOCOURT, Marie-Firmin, (1819-1904), French herpetologist. Worked for the Paris Natural History Museum as a field herpetologist in Mexico, Central America and Thailand, and was partly instrumental in compiling the reptile section of *Mission scientifique au Mexique et dans l'Amerique Centrale: Etudes sur les reptiles et les batraciens* (1870-1909), with Brocchi, Mocquard and Duméril.

BODY CAVITY *See* COELOM.

BODY GROOVE A lateral fold in the body surface of certain reptiles and amphibians.

BODY LENGTH When measuring reptiles, the total length from snout to tail tip. *See* SNOUT-VENT LENGTH.

BOIDAE Pythons and boas. Family of the SQUAMATA, central order Henophidia, inhabiting the tropical and subtropical parts of the world. Over 95 species in some 23 genera in seven subfamilies.

BOJANUS, Ludwig Heinrich, (1776-1827), Lithuanian zoologist at Vilnius University. Produced the significant monograph *Emys, anatome testudinis Europaeae* (1819).

BOLUS A small, soft pellet or lump of faecal matter.

BONN CONVENTION; CMS The Convention on the Conservation of Migratory Species of Wild Animals is an intergovernmental treaty concluded under the aegis of the United Nations Environment Programme and was established with the aim of conserving terrestrial, marine and avian migratory species throughout their range. The Convention's membership includes some 86 Parties, from Europe, Africa, Central and South America, Asia and Oceania, all concerned about the conservation of the world's wildlife and habitats

BONY STYLE The presence of OSSIFICATION in the unpaired, central and hindmost section of the PECTORAL GIRDLE.

BOOIDEA A SUPERFAMILY of snakes consisting of the pythons and boas and the dominant snake species during the TERTIARY period (around 65 to 20 million years ago). Many of these snakes, most of which have long since become extinct, were enormous reptiles up to 15 metres in length. Booid characteristics include a flexible upper jaw, large ventral scales, and an obvious demarcation between head and body. *See also* ANILOIDEA; ACROCHORDOIDEA; COLUBOIDEA.

BOREAL FOREST A type of vegetation occurring in a climatic zone which has cold, snowy winters and short, warm summers, characterized by a flora dominated by hazels and pines.

BOSS An elevated swelling or rounded area; a knob-like protuberance; in certain toads of the BUFONIDAE, a rounded eminence on the mid-line of the head between the eyes, or on or near the tip of the snout.

BOULENGER, George Albert, (1858-1937), Belgian zoologist. Worked in London at the British Museum (Natural History) where he compiled the *Catalogue of the snakes in the British Museum (Natural History)* which was produced in separate volumes (1882-1896). Contributed over 600 papers to many scientific journals. Books include *The tailless batrachians of Europe* (1897 & 1898).

BOURRET, Rene, (1884-1950), French herpetologist. Author of several works on South-East Asian reptiles, including *La Faune de l'Indochine: reptiles* (1927) and *Les Serpentes de l'Indochine* (1936).

BRACHIAL Relating to, or located on, the upper or humeral part of the forelimb; a term for any of the scales on the humeral part of a lizard's forelimb, or BRACHIUM.

BRACHIUM The part of the forelimb containing the HUMERUS.

BRACHYCEPHALIDAE Short-headed toads. Family of the ANURA, inhabiting south-east Brazil. Two species in two genera.

BRACKISH Describing a body of water that is slightly briny or salty. Brackish water is often necessary for the successful development of certain amphibian larvae, as in the natterjack toad *Bufo calamita*.

BRADYARRHYTHMIAS; BRADYCARDIA An abnormally slow heart rate.

BRANCHIA (pl.) **BRANCHIAE** The external gill in aquatic animals, e.g., the larvae of amphibians. Branchial, or branchiate, animals are furnished with gills, or branchiae.

BRANCHIAL Of, or relating to the gills.

BRANCHIAL ARCH; BRANCHIAL BASKET; GILL ARCH In aquatic vertebrates, the skeletal structure that supports the gills.

BRANCHIAL APERATURE *See* SPIRACLE.

BRANCHIAL BASKET *See* BRANCHIAL ARCH; GILL ARCH.

BRANCHIAL FISSURE A single GILL SLIT on the side of the neck in certain aquatic urodeles, such as *Amphiuma*, which are without external gills.

BRANCHIATE; BRANCHIFEROUS Possessing GILLS, or BRANCHIAE.

BRAN MITE *See* MITE.

BRAZIL, Vital Mineiro da Campanha, (1865-1950), South American herpetologist and medical practitioner, founder of the Butantan Snake Serum Institute in Brazil where antivenins against the bites of many venomous snake species were developed.

BRC Acronym denoting the Biological Records Centre (United Kingdom). The BRC is the nationwide guardian of information on the distribution of British wildlife.

BREAKAGE PLANE; AUTOTOMY PLANE; BREAKAGE POINT; BREAKAGE JOINT; FRACTURE PLANE The SEPTUM, or dividing partition of soft tissue that passes through the centre of a tail vertebra and along which separation, or breakage, occurs in AUTOTOMY.

BREAKAGE JOINT; BREAKING JOINT *See* BREAKAGE PLANE.

BREAKAGE POINT *See* BREAKAGE PLANE.

BREATHING PORE *See* SPIRACLE.

BREEDING SIZE The number of individuals in a given population that are occupied in reproduction during a specific generation. It excludes the non-breeding members of the population.

BREHM, Alfred Edmund, (1829-1884), German zoologist. Director of Hamburg Zoo from 1863 and, in 1869, established the Berlin Aquarium where much importance was placed on the amphibian and reptile collections.

BRIDGE The part of the shell in chelonians that joins the upper (CARAPACE) and lower (PLASTRON) sections.

BRILLE The transparent scale that covers the eye in all snakes and some burrowing lizards. Also called SPECTACLE and, less commonly, WATCHGLASS.

BRISTLE *See* TACTILE BRISTLE.

BROAD-SPECTRUM LIGHTING In herpetology, normally used in relation to vivarium lighting equipment that can emit light which is as close as possible to the quality of natural sunlight.

BROAD-SPECTRUM VERMIFUGE Any one of a number of drugs that are extremely effective in expelling or destroying intestinal worms from different groups and, in many cases, their larval stages.

BRONCHOSPASM A constriction of the smooth muscles coating the walls of the bronchi resulting in a whistling or rasping sound during breathing.

BROOD **1.** To sit on, incubate, or hatch, eggs. Used in reference to snakes of the BOIDAE, subfamily Pythoninae, the females of which coil about their eggs during the period of incubation. **2.** To remain with and protectively care for, or cover, eggs. Used in reference to certain lizards and many amphibians, especially salamanders of the PLETHODONTIDAE family.

BROOD CARE The attention given to eggs and young. The different types of brood care seen in reptiles and amphibians are wide and varied and range from the travelling of hundreds of kilometres to find a particular nesting site, as in the marine turtles, and the brooding of the egg clutch, as seen in some python species, to the protection of the nest and eggs, and even assistance in the hatching of the young, in certain crocodilians. *See* BUCCAL INCUBATION.

BROOD POUCH Any pouch, cavity or sac in an animal's body in which eggs can be placed and remain whilst they develop until they hatch. The female Surinam toad *Pipa pipa* has many such brood pouches on her back into which her eggs are embedded. In the females of certain hylid frogs, such as *Gastrotheca*, the brood pouch takes the form of one enlarged sac into which the whole clutch of eggs is placed.

BRUMATION A condition of torpor during extended periods of low temperatures; a state of inactivity during which the metabolic processes are greatly reduced but without actual HIBERNATION.

BUCCAL Of, or relating to, the cheek or mouth.

BUCCAL FORCE PUMP In amphibians, the respiratory system in which air is forced into the lungs by elevating the floor of the mouth while the valvular nostril is shut.

BUCCAL INCUBATION; MOUTH BROODING The eggs of the frog *Rhinoderma darwinii* are deposited by the female on a damp substrate where they remain until they develop. Upon hatching the male then 'swallows' the larvae directing them into specialized vocal sacs. Here the larvae remain, whilst they metamorphose, for a period of some five to six weeks until they are 'regurgitated' by the male as miniature frogs.

BUCKLER A small, round shield; has been used in reference to both the CARAPACE and the PLASTRON of chelonians.

BUFONIDAE True toads. Family of the ANURA, inhabiting most parts of the world except Madagascar, the Arctic and Antarctic regions and New Zealand, as well as many Pacific islands. Introduced into Australia and New Guinea (*Bufo marinus*). Over 200 species in 20 genera.

BUFONIFORM Having the appearance of anurans of the BUFONIDAE family.

BUFOTOXIN One of several TOXINS occurring in secretions from the PAROTOID and many other glands of toads, particularly those of the genus *Bufo*, which act as extremely effective defence mechanisms.

BULL Male; the masculine name for many animals and, in herpetology, used for an adult male crocodilian or chelonian.

BULLA A large blister, normally greater than 20 millimetres in diameter, typically containing serum or blood. *See* BLEB.

BUNDLE In the CHAMAELEONIDAE, the sole anatomical part resulting from the union of the two, or three, digits on each foot thus forming two opposing bundles adapted for clasping.

BURSA A small fluid-filled sac that lessens friction between movable parts of the body particularly at the joints

BUSHMASTER *Lachesis muta*, a PIT VIPER of the family VIPERIDAE, subfamily Crotalinae, inhabiting the tropical rainforests of southern parts of Central America, the southern half of the South American mainland and Trinidad. Although very dangerous it occurs away from human habitation. MONOTYPIC.

BUTTOCK TUBERCLE The enlarged, conical scale, or SPUR, located on the rear upper part of the leg of some chelonians, e.g., *Testudo graeca*.

BUTTON 1. The first and smallest permanent rattle segment acquired by a JUVENILE rattlesnake. As the snake matures the button usually breaks off. At birth, rattlesnakes possess a 'prebutton' which is lost when they have their first shedding, exposing the button. 2. The bony OSTEODERM present in the scales of many crocodilians. Those crocodiles and alligators that have osteoderms are characterized by having single buttons, one osteoderm button per scale, while all caimans have double buttons, two osteoderms per scale.

C

C; °C Celsius; degrees centigrade.

c/ Clutch of (followed by number of eggs e.g, c/6).

cb/cb Captive bred, captive born.

cb/ch Captive bred, captive hatched.

CABLE The jelly-like flexible string in which certain plethodontid salamanders affix their eggs to submerged objects.

CADUCEUS 1. In classical mythology, a rod entwined with two serpents and conveying a pair of wings at its peak, carried by Hermes (Mercury) as messenger of the gods. 2. An emblem similar to this rod and used as an motif of the medical profession. *Compare* STAFF OF AESCULAPIUS.

CADUCIBRANCHIATE Possessing transitory gills; said of an amphibian having gills which are lost as it changes from LARVA to ADULT at METAMORPHOSIS.

CAECILIAN *See* APODA.

CAECILIIDAE Caecilians. Family of the Amphibia, order APODA (or Gymnophiona), inhabiting the tropics, with the exception of Australia and Madagascar. Over 100 species in 26 genera.

CAENOGENESIS The development, in an embryo or larva, of structures and organs that are adaptations to its way of life and are not retained in the adult form.

CAENOPHIDIA An evolutionary group of 'modern' (advanced) snakes that diverged from the HENOPHIDIA. and containing the vipers, sea snakes, elapids, and colubrids, totalling over 3000 species. Caenophidians are thought to be a recent lineage evolving during the mid CENOZOIC period. Unlike the henophidians they lack signs of a pelvic girdle and have no teeth on the premaxillae. Venom is exclusive to caenophidian snakes and, unlike henophidian and scolecophidian snakes, their dorsal scale rows and broad ventral scales are consistent with their vertebrae.

CAENOPHIDIAN *See* CAENOPHIDIA.

CAESAREAN DELIVERY A term used in herpetology in reference to the manual opening of an egg and removal of the embryo, executed only when the safety of the embryo is thought to be at risk, as for example when the embryo of a post-term egg is suspected of being too weak to hatch without assistance.

CAIMAN Any one of a group of tropical American crocodilians of the genera *Paleosuchus, Caiman* and *Melanosuchus*, belonging to the family ALLIGATORIDAE. Similar to alligators but having a more heavily armoured belly.

CALCANEUM; CALCANEUS The largest tarsal bone in vertebrates, corresponding to the heel in man. The elongated calcaneum of anurans is joined at both ends to the astragalus, forming the additional segment in the hind limb.

CALCIUM A soft silvery-white

metallic element that is an essential constituent for all living organisms, being necessary for normal growth and development. Calcium is an important component of teeth and bones and is present in the blood, where it is required for muscle contraction and other metabolic processes.

CALCIUM CYCLE Reproductively active female day geckos of the genus *Phelsuma* go through a calcium cycle which involves the storing of calcium in the ENDOLYMPHATIC SACS, noticeable as whitish swellings on either side of the neck, and its subsequent deposition as egg shell.

CALCULE Any one of a number of small, outward-curving, smooth scales occurring on the back of the Chinese crocodile lizard (*Shinisaurus*).

CALIPASH The greenish glutinous part of the marine turtle found beneath the CARAPACE and considered a delicacy.

CALIPEE The yellow glutinous part of the marine turtle found beneath the PLASTRON and considered a delicacy.

CALL The loud trill, whistle, chirp or song of an anuran. In most species it is only the males which call, although the females of a few species also have a certain VOCALIZATION.

CALLUS A raised area of hard or horny skin; used generally in reference to the lower surfaces of the digits. The TUBERCLE on the foot of an anuran.

CAMBRIAN The earliest period of the PALAEOZOIC, which occurred from approximately 590 million to 510 million years ago, and was characterized by the appearance of algae and many invertebrates, particularly marine forms such as trilobites.

CAMOUFLAGE; CRYPSIS The method by which an animal eludes the detection by predators, typically because of a similarity to their surroundings. *See* CRYPTIC COLORATION; DISRUPTIVE COLORATION; and DISRUPTIVE OUTLINE.

CANALICULUS (pl.) **CANALICULI** A small tubular passage, furrow or groove as in the fangs of snakes.

CANDLING *See* TRANSILLUMINATION.

CANDY The term given to a highly contrasting AMELANISTIC morph. In the corn snake *Pantherophis guttatus guttata* candy specimens have deep reddish-orange blotches on a whitish ground colour.

CANKER; STOMATITIS Commonly called 'mouth rot' and occurring in lizards and, more frequently, snakes; an ulceration particularly of the labial region and oral cavity linings.

CANNIBALISM The act of feeding upon the flesh by an animal of the same species. In reptiles and amphibians cannibalism can result from a number of factors, including overpopulation and/or lack of food. In captivity, if two snakes grasp the same food item the larger will often devour the smaller, although this is usually considered to be accidental; adults also will occasionally look upon their young as prey. Close confinement in vivaria which are too small may encourage such behaviour. A special type of cannibalism is to be found in the tadpoles of certain species of anuran, such as *Hoplophryne*, in which, because they live in water having only a limited supply of food, the majority are devoured by those which are larger and stronger.

CANOPY The layer of often dense vegetation raised high above the ground and usually consisting of tree branches, foliage and epiphytes etc. In tropical rainforests the tree canopy may be over 30 metres above the ground.

CANTHAL Any one, or all, of the scales located along the upper surface of the CANTHAL RIDGE. The canthals lie behind the level of the PRENASAL and POSTNASAL suture and in front of the SUPRAOCULAR. When the canthals are large and in contact along the mid-line, they are referred to as PREFRONTALS.

CANTHAL RIDGE; CANTHUS ROSTRALIS A ridge separating the side of the snout from the top of the head along the line from the tip of the snout to the eye. It may be sharply angular or gently rounded.

CANTHUS The inner or outer corner, or angle, of the eye created by the natural connection of the eyelids.

CANTHUS ROSTRALIS *See* CANTHAL RIDGE.

CANTIL Mexican moccasin *Agkistrodon bilineatus*, a PIT VIPER of the family VIPERIDAE, subfamily Crotalinae, inhabiting swamps, lowland forests and cane fields from southern Mexico southwards through Guatemala, El Salvador, Nicaragua, Honduras and Belize.

CAPSULAR CAVITY; CAPSULAR CHAMBER In amphibians, the minute, fluid-filled space sandwiched between the egg and its surrounding JELLY ENVELOPE and within which a fertilized egg can float freely.

CAPSULAR CHAMBER *See* CAPSULAR CAVITY.

CAPTIVE BREEDING The breeding and rearing of one or more species under controlled conditions.

CAPTIVE DISTURBANCE Psychological and physical disturbances of captive reptiles and amphibians can be directly attributed to specific conditions within their vivaria, or to events experienced by the animals during capture and/or transportation, as when psychological shock or physical damage results in newly caught animals refusing to feed. Incorrect captive conditions can also result in other disturbances, such as vitamin deficiency deformations and stunted growth.

CARAPACE In reptiles of the order Chelonia (turtles, terrapins, tortoises), the domed dorsal part of the shell composed of plate-like bones covered on the outside by horny plates, or SCUTES. In some turtles, such as the soft-shelled (TRIONYCHIDAE), it is completely bony with a layer of skin covering it. The thoracic vertebrae and ribs are included in the carapace, but the girdles of the limbs are separate and situated inside the carapace. The flatter under part of the shell is termed the PLASTRON.

CARBONIFEROUS The second most recent period of the PALAEOZOIC, from approximately 350 million to 280 million years ago, characterized by the appearance on swampy land of amphibians, several primitive reptiles and giant ferns.

CARBUNCLE *See* EGG TOOTH.

CARCINOGEN Any substance which can induce cancer. Certain commercial insecticidal strips used in the eradication of MITES in a captive environment, for example, may be carcinogenic if used carelessly.

CARDIOTOXIN Any poison which attacks the operation of the heart. The venom of the taipan *Oxyuranus scutellatus* is cardiotoxic.

CARETTOCHELYIDAE Pig-nosed turtle. Family of the CHELONIA, suborder Cryptodira, inhabiting southern New Guinea (Fly River), Australia (Daly River). MONOTYPIC.

CAREY *See* TORTOISESHELL

CARINA *See* CARINATE.

CARINATE Having carinas, or keels, as seen in the scales of certain lizards. (Bicarinate: having two keels. Tricarinate: having three keels).

CARNIVORE Any animal (or plant) that feeds on other animals.

CARNIVOROUS Feeding entirely upon the flesh of animals.

CARPAL 1. A bone in the distal region of the forelimb of tetrapods, situated between the RADIUS and/or ULNA and the METACARPALS. **2.** A scale in reptiles situated on the forefoot between the digits and the wrist joint. To indicate scale position the term is often used with a prefix: 'infra-', 'sub-', 'supra-' etc.

CARR Woodland growing in waterlogged conditions with characteristic flora and fauna and often home to a variety of reptiles and amphibians.

CARUNCLE In herpetology, the sharp, horny tubercle on the tip of the snout in baby turtles, crocodilians and rhynchocephalians, used to cut a slit in the egg shell at the time of hatching. The prominence is not a true tooth and differs from the calcified EGG TOOTH found in the hatchlings of snakes and lizards.

CASCABEL; CASCAVEL A, respectively, Spanish or Portuguese name for one of three tropical rattlesnakes of the family VIPERIDAE, subfamily Crotalinae, inhabiting mainland South America from southern Mexico to northern Argentina. *Crotalus simus* inhabits Central America and Mexico and *Crotalus totonacus* northwest Mexico, whereas *Crotalus durissus* is entirely South American. All were formerly *Crotalus durissus*.

CASQUE The helmet-like structure consisting of thickened skin and/or bone on the heads of certain lizards (e.g., CHAMAELEONIDAE; IGUANIDAE), and amphibians (e.g., casque-headed tree frog, *Tetraprion*).

CASTING CYCLE The period between each skin shedding or ECDYSIS.

CASTINGS The undigested, matted remains (bones, hair, feathers etc.) of prey that are regurgitated occasionally by large reptiles such as crocodilians, monitors (VARANIDAE) and large snakes.

CATADONT Having teeth only in the lower jaw.

CATERPILLAR MOVEMENT *See* RECTILINEAR MOVEMENT.

CAUDAL Pertaining to the tail or region of the tail; any pattern, plate, scale or structure on or near the tail of a reptile or amphibian.

CAUDAL DISC The large, blunt, wrinkled, flattened, elliptical or circular region on the end of the tail occurring in certain snake species especially those of the UROPELTIDAE.

CAUDAL FIN A vertically flattened, membranous appendage on the upper and sometimes lower surfaces of the amphibian tail acting as an organ of locomotion and balance when swimming.

CAUDATA Tailed amphibians, also known as the Urodela, containing the newts, salamanders and related forms. An order of the Amphibia, the tailed amphibians are almost entirely confined to the northern hemisphere.

CAUDO-CEPHALIC WAVE During courtship between snakes, when the male and female are lying close to each other side by side, a wave-like horizontal undulation begins near the vent and ripples forward towards the head, becoming more pronounced as the male moves over and on top of the female and creating light contacts between their bodies. The wave is normally produced in the male snake only and appears to act as a stimulus to encourage the female to mate.

CAYMAN An alternative spelling of CAIMAN.

CBSG Acronym denoting the Captive Breeding Specialist Group (IUCN).

CEMENT ORGAN An alternative term for ADHESIVE ORGAN.

CENOZOIC The present geological era that began some 70 million years ago. Subdivided into two periods, the TERTIARY and the QUATERNARY, the Cenozoic is characterized by the rise of modern organisms, the flowering plants and mammals in particular.

CENTRAL Used in reference to any one of the row of large laminae (usually five) running unpaired down the centre of the CARAPACE in chelonians.

CENTROLENIDAE Glass frogs. Family of the ANURA, suborder Procoela, inhabiting the cloud and rainforests of Ecuador and Columbia in tropical America. 65 species in three genera.

CEPHALIC PLATE Any one of the enlarged scales on the top of the head in reptiles.

CERVICAL Of, or relating to, the neck or cervix; an alternative name for the NUCHAL plate in crocodilians.

CESTODA Tapeworms. Class of flatworms (Phylum: Platyhelminthes; order: Pseudophyllidea) which are parasitic as adults and larvae in both amphibians and reptiles. Cestodes lack a gut and their bodies are divided into many egg-producing segments.

CESTODE *See* CESTODA

CHAMAELEONIDAE Chameleons. Family of the SQUAMATA, suborder SAURIA, inhabiting Africa, Madagascar, south Asia and southern Europe. Over 85 species in five genera: *Brookesia*, *Calumna*, *Chameleo*, *Furcifer*, and *Rampholeon*. (Occasionally spelt Chamaeleontidae).

CHARACTER DISPLACEMENT The belief that two species are more dissimilar where they occur together (i.e., are SYMPATRIC) than where they are unconnected geographically (i.e., are ALLOPATRIC).

CHECKLIST A list of species from a certain area or taxonomic group, or both, sometimes including identification keys, with space provided so they can be marked off as they are seen.

CHELIDAE Side-necked turtles. Family of the CHELONIA, suborder Pleurodira, inhabiting Australia, New Guinea and South America. 40 species in nine genera.

CHELONIA The order of some 250 species of reptiles, containing the terrestrial tortoises, marine turtles and freshwater terrapins. The body is encompassed by a bony-plated, horny, scale-covered shell with a CARAPACE above and a PLASTRON below, and the horny jaws lack teeth The order is divided into two families: CHELONIIDAE (the sea turtles), and TESTUDINIDAE (the freshwater turtles and land tortoises).

CHELONIIDAE Sea turtles. Family of the CHELONIA, suborder Cryptodira, inhabiting much of the world's oceans from the tips of the southern continents northwards almost to Scandinavia. Eight species in four genera.

CHELYDRIDAE Snapping turtles. Family of the CHELONIA, suborder Cryptodira, inhabiting southern Canada into the eastern USA. Two MONOTYPIC genera.

CHERNOV, Sergei Alexandrovitch 1903-1964, Russian herpetologist at the Academy of Science's Zoological Museum in Leningrad. Contributed papers on the Tadzhikistan and Armenian reptiles and amphibians and worked on the amphibian section in *Animal world of the USSR* (1936-1953). With Terent'ev, he co-authored the *Key to the reptiles and amphibians of the USSR* (1949).

CHICKEN The name commonly applied to a juvenile green turtle *Chelonia mydas*.

CHIGGER; CHIGOE; REDBUG The name for any one of the tiny red-coloured mites of the order ACARINA, suborder Trombidiformes. Chiggers, or harvestmites, are free-living parasites and both adults and larvae of some species often infest reptiles, particularly lizards, and can cause severe itching in humans. The chigger mites of the Orient transmit scrub typhus.

CHIGOE Although this is the common name for the tropical flea, *Tunga penetrans*, the female of which lives on or tunnels into the skin of its host which includes man, the name is also used occasionally for the CHIGGER mite.

CHINSHIELD Any one of the large, paired, elongated scales situated immediately behind the first pair of infralabials on either side of the mid-line on the lower jaw of snakes. Chinshields are always in pairs, with a single pair most often occurring in snakes. In lizards they are the series of paired scales on the mid-line of the lower jaw, behind the MENTAL scales.

CHLAMYDIAL INFECTION; CHLAMYDIOSIS Chlamydiae are a group of microorganisms, in size midway between bacteria and viruses. They are responsible for a variety of transmittable diseases in humans and animals especially birds and, in recent years, outbreaks of chlamydiosis have occurred in mainly captive but also some wild amphibians causing average to high mortality rates in several species including *Xenopus laevis*, *Myxophyes iteratus*, and *Ceratobatrachus guentheri*. As with viruses, chlamydiae can only multiply by first invading the cells of another life-form.

CHOANA (pl.) **CHOANAE** A funnel-like opening. Used in reference to the hindmost internal openings of the nasal passages in the roof of the mouth. *See* NARIS.

CHOLINERGIC Pertaining to nerve endings which liberate acetylcholine.

CHORDATA A large PHYLUM consisting of the animals that have a NOTOCHORD which, in higher forms, is protected by a vertebral column. In addition to the amphibians and reptiles, the phylum also includes the fish, birds, and mammals. The first CHORDATES and the earliest vertebrates are found in CAMBRIAN rocks.

CHORDATE An animal of the phylum CHORDATA, possessing a NOTOCHORD, a dorsal hollow nerve cord (a fluid-filled tube that extends the length of the body), PHARYNGEAL gill slits or pouches, and a tail at some stage in its life.

CHORIO-ALLANTOIC PLACENTA The membrane caused by the uniting of the ALLANTOIS and the CHORION in reptiles having VIVIPARITY. The membrane acts as a form of placenta by allowing a certain amount of exchange between the circulations of the adult and the FOETUS.

CHORION One of the three membranes that are formed during the embryo's development within the egg, and which envelops and protects both yolk sac and embryo.

CHOROID In vertebrates, a layer of tissue in the eye lying directly outside the retina and containing blood vessels and pigment.

CHORUS An gathering of 'singing' male frogs.

CHROMATOPHORE A pigment cell of the skin; a specialized body cell that contains the pigment granules which largely decides the coloration of an animal. By diffusing or constricting such granules some animals are able to change their colour.

CHROMOMYCOSIS An infection, occurring in anurans, caused by any one of a variety of Ascomycota pigmented fungi species (e.g., *Phialophora*, *Scolecobasidium*, *Cladosporium*).

CHUCKWALLA *Sauromalus obesus*, a lizard of the family IGUANIDAE inhabiting rocky hillsides and lava flows on several islands and the mainland of west and south western North America and Mexico.

CHYTRID FUNGUS DISEASE *See* CHYTRIDIOMYCOSIS.

CHYTRIDIOMYCOSIS An epidemic disease the source of which is a species of primitive, parasitic, chytrid fungus *Batrachochytrium dendrohatidis*, found in damp soils and freshwater habitats. Infecting the exterior, keratinized layer of skin in a range of frog and toad species, the disease has been responsible for up to 100% mortality in the populations of several species. Chytridiomycosis is linked with the recent decline in populations of rainforest amphibian species in Australia and Central America. and, at present, seems to be eliminating other species as it spreads around the world

CILIARY Any one of the small scales that border the edge of the eyelid in crocodilians and lizards. When describing the upper and lower ciliaries the prefixes 'supra-' and 'infra-' are used respectively.

CIRCADIAN RHYTHM A cycle of behaviour and biological processes that occurs regularly at 24-hour intervals in animals that are normally influenced by internal 'clocks', but which is synchronized to time-related factors from their surroundings. Well defined in many reptiles and amphibians. *See* PHOTOPERIOD.

CIRCUMNARIAL *See* CIRCUMNASAL.

CIRCUMNASAL; CIRCUMNARIAL In the CROCODILIA, any one of the row of scales that encircle the nasal protuberance. Also used to describe the enlarged scales, other than the ROSTRAL and LABIAL scales, surrounding the nostril in the geckos (GEKKONIDAE).

CIRCUMTROPICAL Around the world situated between the two tropics (23° 30' N and 23° 30' S).

CIRRUS (pl.) **CIRRI** Any one of the projections which descend from the nostrils in the males of some lungless salamanders. The NASOLABIAL GROOVE reaches downward close to the tip of each cirrus.

CITES Acronym denoting the Convention on International Trade in Endangered Species of Wild Fauna and Flora (also known as the

41

Washington Convention). CITES is an international treaty, convened in 1973 in Washington, D.C. and becoming effective on July 1, 1975. The United States and some 112 other nations are party to this agreement and meet and discuss wildlife trade issues at least once every two years. CITES recognises that unlimited commercial exploitation is a key threat to species' survival and establishes worldwide controls over trade in certain species of endangered and threatened animals and plants.

CLADE In CLADISTICS, a family branch resulting from a split in an earlier ancestry. Such a separation produces two distinctive new taxa, each of which is represented as a branch in a phylogenetic diagram, or CLADOGRAM.

CLADISTICS The use of PHYLOGENETIC SYSTEMATICS to produce a taxonomic system that is employed in the study of evolutionary associations.

CLADOGENESIS In CLADISTICS, the evolution of a new TAXON as a result of the splitting of a family lineage to produce two equivalent sister groups, taxonomically detached from the ancestral taxon. *See* ADAPTIVE RADIATION.

CLADOGRAM A chart that defines the branching progression in an evolutionary tree.

CLASPING REFLEX The voluntary contractive and spasmodic grasping of any object by a male anuran when touched on the underside of the forelimb or in the region of the chest,

during periods of mating activity. The European common toad *Bufo bufo*, for example, will frequently clasp the fingers of a person if a hand is slowly lowered into the water amongst males awaiting female mates. Not only does it aid the anurans in sex recognition, it also serves to guarantee a male's presence when the female deposits her eggs.

CLASS The category in taxonomy which ranks between the PHYLUM and the ORDER. All reptiles are placed in the class REPTILIA and all amphibians in the class AMPHIBIA.

CLASSIFICATION The cataloguing of living things into systematic groups. *See* TAXONOMY.

CLAW A continually growing curved, pointed, horny process on the end of each digit in certain reptiles, and also birds and some mammals.

CLAVATE Shaped like a club; with the thicker end uppermost.

CLAVICLE A bone that forms part of the PECTORAL (shoulder) GIRDLE linking the SCAPULA to the STERNUM. The clavicle is generally quite prominent in *Sphenodon*, lizards and anurans, but is frequently much reduced or absent altogether.

CLEIDOIC EGG An egg having a tough, protective shell which allows gaseous exchange but which restricts water loss (although water may be absorbed). Cleidoic eggs are characteristic of reptiles exhibiting OVIPARITY.

CLIMAX PERIOD In anuran METAMORPHOSIS, the stage in

the development of the LARVA from the beginning of the emergence of the forelimb to the development of the mouth and TYMPANUM and the resorption of the tail.

CLINAL SPECIATION A form of ALLOPATRIC speciation in which, when a geographical barrier falls across a CLINE, a species, already displaying some variation, is divided into two parts but continue their divergence in their new separation.

CLINE A continuous and gradual change of a variable characteristic between members of a species over a given ecological or geographical range.

CLITORIS In certain reptiles, an erectile rod of tissue that is the female equivalent of the male penis.

CLITOROPENIS *See* PENOCLITORIS.

CLOACA (pl.) **CLOACAE** The common chamber into which the genital, urinary and digestive canals release their contents, and which opens to the exterior through the anus or VENT. In the ARCHOSAURIANS (dinosaurs, crocodilians and birds), the cloaca is parallel to the axis of the body. In the LEPIDOSAURIANS (amphisbaenians, snakes, lizards and tuataras), the cloaca is transverse to the body axis.

CLOACAL BONES A pair of small bones, each of which lies in a cavity directly behind the VENT, on either side of the tail, in geckos and snake lizards (GEKKONIDAE and PYGOPODIDAE).

CLOACAL CAPSULE The thickened cloacal wall in females of certain snake species in which the HEMIPENIS of the males is densely covered with spines. The cloacal capsule can be felt under gentle pressure as a small, firm lump and is therefore useful in determining the sex of an individual.

CLOACAL GAPING The sexual activity occurring in the females of certain snake species consisting of raising the tail and everything the cloaca indicating receptivity to males.

CLOACAL GLAND 1. The MUSK GLAND of crocodilians and certain chelonians **2.** Part of a large, cylindrically-shaped gland surrounding the opening to the anus in male urodeles. The gland discharges into the CLOACA via PAPILLAE lining the cloacal wall.

CLOACAL LIP *See* CLOACAL LABIA.

CLOACAL LABIA; CLOACAL LIP In the URODELA, the swollen ridge, particularly noticeable in breeding adults, situated on either side of the VENT.

CLOACAL PLATE *See* ANAL PLATE.

CLOACAL POPPING A form of defensive action employed by a few North American snakes (e.g., the western hooknosed snake *Ficimia cana*, and the coral snakes *Micruroides euryxanthus* and *Micrurus fulvius,* in which the CLOACA is rapidly extruded and withdrawn, resulting in an audible bubbling or popping sound whilst the writhing tail is held aloft.

CLOACAL PROBE Any tapered, blunt, smooth instrument inserted caudally, after lubrication with a non-spermicidal fluid such as tap water or sterile saline, into the CLOACA of a snake in order to determine its sex. The sex of a snake can often only be determined with any certainty by the insertion of such a probe into either of the two LUMINA at the base of the tail.

CLOACAL SPUR *See* ANAL SPUR.

CLOACAL SWELLING In urodeles, the swollen region around the vent, especially noticeable in breeding adults.

CLOACITIS A condition seen in terrestrial chelonians manifesting itself as a malodorous, obnoxious leak or discharge from the tail. Veterinary diagnosis is essential as the condition could indicate a flagellate infection.

CLONIC The alternating contraction and relaxation of muscles.

CLOTTING TIME The amount of time taken for whole blood to clot, generally four to eight minutes.

CLOUD FOREST A montane forest characterized by frequent heavy mists, high humidity, cool temperatures and dense, luxurious plant life, normally occurring on the windward slopes of tropical mountains and usually at heights above 1000 metres.

CLUTCH 1. The collective noun for the full complement of eggs laid by a single female at one time. The average clutch size of some snakes (e.g. pythons) and lizards (e.g. green iguanas) may exceed 50 in number. **2.** Egg production and deposition. Females of certain species may clutch several times a year.

CLUTCH MATE Used in reference to any one of the individuals hatched from the same CLUTCH of eggs.

CMS Acronym denoting the Convention on the Conservation of Migratory Species of Wild Animals. Also known as the BONN CONVENTION.

CNEMIAL In crocodilians, any one of the large and sometimes flap-like scales that project on the hindmost fringe of the fore- and hind limbs.

COACHWHIP *Masticophis flagellum,* a snake of the family COLUBRIDAE inhabiting a wide variety of habitats, including prairie, desert, woodland and farmland, in the southern parts of North America, and Mexico.

COAGULANT Or procoagulant. An effect of certain venoms which promotes clotting of the blood within the blood vessels. This rapidly uses up all of the clotting factor in the blood and leads to prolonged bleeding similar to that caused by ANTICOAGULANT. Early signs include continual bleeding from the site of the bite and from any razor cuts to the bitten limb which may lead to excessive blood loss and are potentially dangerous. Taipan (*Oxyuranus*) and brownsnake (*Pseudonaja*) venoms are strongly coagulant.

COAGULATION The formation of a blood clot from liquid blood.

COAGULOPATHY In relation to

certain venomous SNAKE BITES, a disruption in the blood clotting process resulting from the reduction of clotting factors such as fibrinogen.

COCCIDIOSIS A protozoan disease that affects cells of the alimentary canal and the blood, most often in juveniles, particularly those of chameleons and snakes; caused by parasitic organisms of the order Coccidia.

COCCYGEAL STRIPE A stripe running longitudinally across the centre of the rump in some frogs and normally referring to a stripe that only crosses that area of the vertebral column to the rear of the SACRUM.

COCHLEA Part of the inner ear of certain reptiles that converts sound waves into nerve impulses.

COCHRAN, Doris M., (1898-1968), American herpetologist, Curator of the Division of Reptiles and Amphibians of the United States National Museum, Smithsonian Institution in Washington DC. Published some 80 scientific and popular articles and described five new genera and 90 new reptile and amphibian species. Other publications include *Poisonous reptiles of the world: A wartime handbook* (1943), *Living amphibians of the world* (1961), and *The new field book of reptiles and amphibians* (1970).

COCKATRICE 1. A legendary monster that was part cockerel and part snake and which had the ability to kill with just a momentary look. **2.** An alternative name for BASILISK.

COELOM A fluid-filled cavity within the body of an animal and usually referring to a cavity lined with a specialized tissue PERITONEUM in which the gut is suspended. Snakes, lacking a diaphragm, have no abdomen but instead possess a fluid-filled abdominal cavity, or coelom. The presence of a coelom allows separation of the gut from the body wall, providing greater mobility and specialization, and requiring the development of a blood-vascular system. It contains the digestive tract and visceral organs and its structure and development is an important feature in identifying major groups of animals.

COGBAZ Acronym denoting the Council of Governing Bodies of Australasian Zoos.

CO-HOUSING The keeping of males and females of the same species, or individuals of different species, together within the same enclosure.

COIL A loop, or sequence of loops, in a constricting snake's body and thrown about its prey in order to subdue it.

COITUS A technical term for copulation, or sexual union.

COLD-BLOODED At one time used to describe all amphibians and reptiles and, although still used today, the term is now generally considered to be unsuitable, particularly for reptiles many of which may sometimes have body temperatures far above that of so-called 'warm-blooded' animals inhabiting the same region. *See*

HOMOIOTHERM; POIKILOTHERM.

COLLAGENOLYTIC The collapse of fibrous proteins normally occurring in the skin, bones, ligaments and cartilage.

COLLAR 1. A transverse fold of skin on the under surface of the neck in certain lizards, normally covered with enlarged scales. *See* GULAR FOLD. **2.** A band of colour across the nape of the neck, as in the North American collared lizard *Crotaphytus collaris*.

COLONY 1. A group of the same species of animal, or plant, living together. **2.** A collective noun for a group of group of frogs.

COLORATION Fixed chiefly by chromatophores in the skin. Reptiles and amphibians generally possess only yellow, red and brown-black colour pigments. Varying shades of blue and green are caused via layering, refraction and scattering of light. Animal coloration often acts as camouflage by softening and breaking up the hard outline of the contours of the body, and gives a certain degree of protection against ultraviolet rays from the sun. Striking and bright colours are frequently found in many poisonous amphibians and venomous reptiles.

COLUBRID Any snake of the family COLUBRIDAE. Although the majority are harmless to man, many species possess enlarged teeth on the rear of the MAXILLA, or upper jaw, and a few species produce venom which is strong enough to cause death in man. *See* OPISTOGLYPH.

COLUBRIDAE Typical snakes. Family of the SQUAMATA, central order CAENOPHIDIA, inhabiting every geographical region. Over 2000 species in some 320 genera.

COLUBRINE The rounded head of a COLUBRID or ELAPID snake may be said to be colubrine in shape.

COLUBROIDEA A SUPERFAMILY containing the major group of snakes to develop venom and venom apparatus some 40 million years ago. *See also* ACROCHORDOIDEA; BOOIDEA.

COLUMELLA AURIS A bony or cartilaginous rod in reptiles and amphibians (and birds) linking the tympanic membrane to the inner ear and transmitting sound.

COMB 1. In desert lizards, a fringe consisting of numerous slender, elongated scales on the lateral border of the toes.
2. A term given to the FEMORAL PORES of male lizards in breeding condition, which may exude a wax-like substance and which, when viewed from below, give the appearance of teeth on a comb.

COMBAT DANCE A ritualistic 'fight' between two (or more) male snakes in order to establish sexual supremacy over a female.

COMBINATION In nomenclature the name of a SPECIES which is formed by the generic name followed by the specific name, or a SUBSPECIES which is formed by the generic, specific and subspecific names.

COMMON NAME *See* VERNACULAR NAME.

COMPASS ORIENTATION The capability to head in a specific compass direction without the need to refer to landmarks.

COMPENSATION MOVEMENT The condition, seen in amphibians and reptiles, in which the head faces an imaginary, fixed point as the body of the animal is moved. A toad, if placed upon a slowly revolving turntable, will move its head in the opposite direction to which it is being revolved. As the strain on its neck becomes too much, it will shift its position, enabling it continually to fix its gaze on the same point.

COMPETITION A requirement by two or more individuals for the same resource, e.g., mates, territory, food etc.

COMPRESSED In reference to body shape, flattened from side to side providing a greater depth than width. *Compare* DEPRESSED.

CONCAVE Curving inwards like the inside of a sphere; hollow. In herpetology, used in reference to the PLASTRON of certain male chelonians. *Compare* CONVEX.

CONCEALED SURFACE Any body surface which is normally hidden when an animal is at rest. These surfaces are frequently marked with striking colours which are suddenly revealed when the animal is alarmed or aroused, thus startling the intruder.

CONCERTINA MOVEMENT A type of locomotion occurring in snakes in which the animal progresses via a series of alternate contractions and extensions of the body, giving it an accordion-like appearance as it moves across a surface. The movement is similar to that of a looper caterpillar (family Geometridae), but the folding of the body takes place in the horizontal plane rather than the vertical.

CONGENERIC Of the same genus. Species placed in a particular genus are called 'congeners'.

CONGENITAL Present at birth. The term describes, or relates to, any defect, deformity, or abnormal condition existing at birth. The occurrence of some congenital abnormalities, such as cleft palate, is determined by environmental as well as hereditary factors.

CONGREGATION A collective noun for a group of juvenile crocodilians.

CONICAL Descriptive of an elevated scale or LAMINA that narrows to a pointed centre.

CONJUNCTIVITIS An infection of the conjunctiva of the eyes but, in herpetology, normally used to specify a more generalized eye infection, often revealed by an eye discharge.

CONSERVATION The artificial management of the environment's natural biological relationships in order to preserve a particular balance among the species present.

CONSPECIFIC Pertaining to individuals or populations of animals belonging to the same species. Subspecies of the same species are said to show conspecificity.

CONSTRICTION The method employed by some snakes to kill

their prey in which the victim is squeezed and suffocated.

CONSTRICTOR In general, any of the giant snakes, such as the pythons and boas, but used to describe any snake which squeezes its prey to subdue or kill it.

CONSUMER Any organism which must devour other organisms (alive or dead) in order to satisfy its energy requirements.

CONVERGENCE The state in which two unrelated species or groups, in the course of evolution, have developed a superficial resemblance to each other or similar adaptations to their environment, even though they are not closely related and may be widely separated geographically. The situation (convergent evolution) occurs when the conditions in which the animals live are similar.

CONVERGENT EVOLUTION *See* CONVERGENCE.

CONVEX Curving outwards like the outside of a sphere. Used, in herpetology, when referring to the CARAPACE of chelonians. *See* CONCAVE.

COOTER; SLIDER Any one of a variety of small North American freshwater turtles of the family EMYDIDAE, such as *Chrysemys concinna* for example, inhabiting rivers, ditches and the mouths of rivers etc..

COPE, Edward Drinker, (1840-1897), American herpetologist, prolific author of over 1260 publications including *The Batrachia of North America* (1889), and *The crocodilians, lizards and snakes of North America* (1900). The American Society of Ichthyologists and Herpetologists named their journal *Copeia* in honour of his important work.

COPPERHEAD *Agkistrodon contortrix,* a PIT VIPER of the family VIPERIDAE, subfamily Crotalinae, inhabiting rocky, wooded areas in parts of North America

COPULATION The act of sexual union, also called 'coition', which leads to the fertilization of the egg by the male sperm.

COPULATORY PLUG *See* SEMINAL PLUG.

CORACOID Either one of the pair of cartilage bones that, in certain reptiles, form the ventral side of the PECTORAL GIRDLE and which, with both scapulae, form the articulation surfaces for the forelimbs.

CORDON A term for an individual string of eggs in toads of the genus *Bufo.*

CORDYLIDAE Girdle-tailed lizards, plated lizards and related species. Family of the SQUAMATA, suborder SAURIA, inhabiting Africa and Madagascar. Approximately 60 species in ten genera.

CORE AREA That part of an animal's range in which it feels sufficiently safe and secure in order to rest, eat, nest or give birth etc. Such core areas may be strongly defended against others of the same species which do not share the range.

CORIACEOUS Of, or resembling, leather and often used as a descriptive of the eggs of many reptiles and the skins of soft-shelled turtles (TRIONYCHIDAE).

CORN (pl.) **CORNUA** *See* HIDE.

CORNEAL LAYER *See* STRATUM CORNEUM.

CORNIFIED *See* KERATINOUS.

CORN-MARK In the commercial skin trade, any one of the protuberances on the inner, or under, surface of the skin of an alligator. These projections, which are horny tissue rooted in the centre of each of the reptile's large DORSAL scutes, render the consequently unattractive hide commercially useless.

CORNUATE Said of an animal having one or more horns or horn-like structures. Lizards of the genus *Phrynosoma* (the horned 'toads') are cornuate.

CORONOID A bone in the lower jaw of some reptiles.

CORRIDOR DISPERSAL ROUTE As formerly defined in 1940 by the American palaeontologist G. G. Simpson, a corridor is a MIGRATION ROUTE that permits more or less unconstrained faunal exchange. As a consequence many or most of the animals of one faunal region can migrate to another one. Such a dispersal corridor has long existed between Western Europe and China via central Asia. *See* FILTER DISPERSAL ROUTE; SEASONAL MOVEMENT; SWEEPSTAKE DISPERSAL ROUTE.

CORYPHODONT Having teeth that increase in length from the front of the mouth to the rear.

CORYTOPHANIDAE The basilisks etc. Family of the SQUAMATA, suborder SAURIA, inhabiting Africa and Madagascar. Some seven species in three genera: *Basiliscus*, *Corytophanes*, and *Laemanctus*.

COSMOPOLITAN Distributed throughout every ZOOGEOGRAPHICAL REGION of the world.

COSTAL A plate on the CARAPACE of a chelonian situated between the VERTEBRAL and MARGINAL laminae and covering the PLEURAL bones.

COSTAL GROOVE Any one of several grooves on the flanks of many urodeles. The area of skin between each groove is called the costal fold.

COTTONMOUTH The North American PIT VIPER *Agkistrodon piscivorus*. The name 'cottonmouth' is derived from the brightly coloured interior of the snake's mouth which is displayed as a means of intimidation. *See* MOCCASIN.

COUNTERSUNK Sunk below the margins of; e.g., the lower jaw of many burrowing snakes which fits securely within the margins of the upper jaw.

COURTSHIP The establishment of pair-bonding, accomplished by various types of display and largely genetically fixed behaviour, which serve to break down aggression between the sexes and encourage sexual interest. Courtship in amphibians and reptiles is extremely varied and contains specific and ritualistic displays associated with both special coloration and body decoration, as well as with stimulation from touch, smell and sound.

COW A sexually mature female terrapin, turtle or crocodilian.

COWL A hood, particularly a loose-fitting one, like that of a monk's habit. Used in some literature for the hood of a cobra.

CRANIAL Of, or relating to, the skull.

CRANIAL CLUB *See* CRANIAL CREST.

CRANIAL CREST A raised ridge on the top of the head of certain toads, between the eyes (interorbital) or behind the eyes (postorbital). In the toad *Bufo terrestris* the cranial crest is enlarged and swollen and called the cranial 'club'.

CRANIAL NERVE Any one of a group of nerves originating in the brain stem and providing a motor and sensory function to the face, eyes, tongue and diaphragm.

CRANIUM The bony part of the skull which encloses the brain.

CRENATE Having a scalloped margin.

CRENULATE Having a finely notched margin.

CREPUSCULAR Active at dusk and/or just before dawn.

CRES Acronym denoting the Centre for Reproduction of Endangered Species (Zoological Society of San Diego Zoo).

CREST Any elevated, flexible, cutaneous ridge or fold on the tails and/or backs of many lizards and which, in the breeding season, develops on the tails and/or backs of male urodeles.

CRESTAL Any one of a number of enlarged, convex and sharply keeled scales which shape the dorsal crests of the Chinese crocodile lizard (*Shinisaurus*).

CRETACEOUS The final and most recent period of the MESOZOIC, from 135 million to 65 million years ago, marked by the continued domination of both land and sea by the dinosaurs until their rapid and widespread extinction at the end of that period. Aquatic and flying reptiles also became extinct during the later part of the Cretaceous.

CREW Acronym denoting the Centre for Reproduction of Endangered Wildlife (Cincinnati Zoo).

CRIBO; YELLOW-TAILED CRIBO *Drymarchon corais corais*, a large robust snake of the family COLUBRIDAE, inhabiting the savannahs, woodlands and swamps of Trinidad, Tobago, Surinam, and Venezuela south through Amazonia and Paraguay to Argentina. Several other cribos occur throughout the tropics including the red-tailed cribo *Drymarchon corais rubidus*, and the grey cribo *Drymarchon corais melanurus*.

CRIMSON *See* FLAME.

CRITICAL MAXIMUM In herpetology, the highest body temperature of amphibians and reptiles which they can endure before death ensues. *See* OPTIMUM TEMPERATURE RANGE; VOLUNTARY MAXIMUM.

CRITICAL MINIMUM In herpetology, the lowest body

temperature of amphibians and reptiles at which locomotion becomes impossible. *See* OPTIMUM TEMPERATURE RANGE; VOLUNTARY MINIMUM.

CROAK The low, hoarse, raucous sound made by male anurans. The sound is normally made with the mouth closed whilst air is forced in and out of the lungs and into the VOCAL SAC or sacs.

CROAK REFLEX A particular response in male anurans when grasped by other males, in which the one grasped utters a low warbling croak which causes the sides of the body to quiver rapidly, making the other release its hold. If the individual grasped is a female no croak is uttered, the hold is continued and spawning takes place.

CROCODILIA The order of large, aquatic, tropical reptiles (subclass ARCHOSAURIA) containing the crocodiles, caimans, alligators and gharials. First recorded from TRIASSIC rocks, they are closely related to the dinosaurs and pterosaurs and, apart from the birds, are the only archosaurs to have survived the MESOZOIC Era. Crocodilians possess long snouts and large, strong conical teeth, a secondary palate which permits breathing in water whilst the mouth is open, short but robust limbs, an armoured skin, and a laterally flattened tail adapted for swimming.

CROCODILIDAE Family of the CROCODILIA inhabiting the tropics and found in every continent except Europe. 13 species in three genera.

CROSSBAND Used to describe an area of colour crossing the body from side to side across the back and reaching down to the VENTER but, unlike a RING, not actually joining.

CROSS-REACTIVITY In reference to the effects and treatment of certain venomous snake bites, the effect of an antivenom developed from the venom of one snake species and used to treat the bite of a different snake species.

CROTALID; CROTAL Any snake of the family CROTALIDAE, or pit vipers, which includes the rattlesnakes.

CROTALINE MOVEMENT *See* SIDWINDING MOVEMENT.

CROTAPHYTIDAE Collared and leopard lizards. FAMILY of the SQUAMATA, suborder SAURIA, and formerly of the IGUANIDAE, inhabiting the south-western states of North America from south-eastern Oregon to the Sonoran and Mojave deserts of Baja California, eastward to central Texas and eastern Iowa, and south to northern Zacatecas, Mexico. 12 species in two genera.

CRUCIFORM Shaped like a cross.

CRURAL Of, pertaining to, or located on or near the CRUS.

CRUS That section of the hind limb between the knee and the ankle containing the FIBULA and the TIBIA.

CRYPSIS *See* CAMOUFLAGE.

CRYPTIC COLORATION The coloration of an animal that blends with its surroundings, providing it with camouflage and enabling it to remain hidden amongst the vegetation or upon the ground on which it rests.

CRYPTOBRANCHIDAE Hellbenders, giant salamanders. Family of the CAUDATA, inhabiting the south-eastern Palaearctic and eastern Nearctic. Three species in two genera.

CRYPTODIROUS Descriptive of chelonians that are able to retract the head into the CARAPACE by means of an S-shaped bend in the vertebral column and pertaining to any member of the suborder Cryptodira.

CRYPTOSPORIDIOSIS An important protozoal disease occurring in lizards and snakes, caused by an internal parasite of the genus *Cryptosporidium*. The parasite, a type of coccidian, has a direct life cycle and can pass directly from one host to another. Symptoms include weight loss and mucous-covered regurgitated food. Untreated, death usually follows in a matter of months.

CRYPTOZOIC Inhabiting secluded and/or dark places, beneath stones, logs and other debris, or in holes and caves.

CUIRASS The hard, outer protective armour of bony shields and plates in crocodilians.

CUNICULINE Dwelling in burrows.

CURAREMIMETIC TOXIN A powerful component (e.g., as little as 4 millionths of a gram may kill a 20g mouse) produced in large concentrations in certain venoms, such as those of sea snakes (*Laticauda* sp.). Death from respiratory failure may result when curaremimetic toxins inhibit the transmission of nerve signals to the skeletal muscles (including the diaphragm) which remain paralysed.

CUSP A tooth like projection, as on the jaw of a chelonian.

CUSPID Possessing tooth like projections. *See* BICUSPID; TRICUSPID.

CUTANEOUS Pertaining to the cutis, or skin. *See* INTEGUMENT.

CUTANEOUS RESPIRATION Breathing through the skin. In the soft-shelled turtles (TRIONYCHIDAE) and amphibia, the surface of the body has become greatly vascularized for gaseous exchange. Such exchange is vital for these animals in which mucous skin glands maintain a moist respiratory surface.

CYANOMORPH A colour phase (MORPH) in which the dominant colour is blue.

CYANOSIS In reference to the effects of certain venomous SNAKE BITES, a bluish discoloration of the skin resulting from an inadequate supply of oxygen to the blood.

CYANOTIC Descriptive of the slight bluish coloration resulting from abnormal amounts of reduced HAEMOGLOBIN in the blood.

CYCLE; CYCLING The recurring reproductive stage, activated by hormonal changes, in turn prompted by environmental cues (ABIOTIC FACTORS). Cycling may take place once, or several times, a year.

CYCLOID Descriptive of a reptile scale possessing an evenly curved, free border.

CYCLOPIA A birth defect in which only one eye is present, often situated in an abnormal position.

CYCLOTREME Having a rounded anal orifice, a characteristic of crocodilians and chelonians.

CYST Usually, any abnormal membranous sac or blister-like pouch containing semi-solid matter or fluid, but also sometimes referring to the thick-walled protective membrane enclosing an organism or cell larvae, e.g., the larvae of some molluscs encyst in tadpoles.

CYTOLYSIS The suspension or destruction of living cells.

CYTOTOXIC *See* CYTOTOXIN.

CYTOTOXIN; PROTEOLYTIC A broad term denoting a substance which can poison or destroy cells; a venom or component of a venom which causes the destruction of tissue resulting in necrosis, or tissue death. Many vipers, especially the puff adder *Bitis arietans*, and terrestrial elapids such as the spitting cobras *Naja mossambica*, *Naja nigricollis* and *Naja pallida*, possess cytotoxic venoms.

D

DABOIA; TICH PALONGA; TIC POLONGA The nocturnal and highly venomous Russell's viper *Daboia russelli* of the family VIPERIDAE found throughout Pakistan, India and south east Asia.

DAGGER An enlarged, sharp, spine like projection on the inside of the forefoot of certain anurans. It is essentially a long pointed bone, situated where the thumb (absent in most anurans) should be and covered only by skin. In the frog *Babina holsti*, the daggers serve as effective weapons and are capable of drawing blood from a human hand if the animal is handled carelessly. The structure is also found in the genus *Petropedates* and, besides defence, may also assist the male to grasp a female at the time of mating.

DAMAGE FIGHT In contrast to RITUAL COMBAT, a fight resulting in severe to fatal injuries to one, or sometimes both animals. In a community of captive lizards such damage fights can occur over the establishment and/or defence of an individual's territory within the vivarium, or during courtship.

DAMMING PHENOMENA The change in water pressure as a result of a shock wave from a moving object and, in amphibians, detected by the LATERAL LINE SYSTEM.

DANCE In reference to snake-charming, the swaying, side-to-side movements of an alert cobra in which the head and front part of the body is raised from the ground and the hood spread. The cobra 'dances' in response to the charmer's rhythmical body movements and not the music, as is often supposed by onlookers.

DANGEROUS WILD ANIMALS ACT The Dangerous Wild Animals Act was originally introduced in the United Kingdom in 1976 as a private members bill in reply to public unease regarding private individuals keeping dangerous wild animals, such as crocodilians and venomous snakes, as pets. The Act aims to make certain that where private individuals keep dangerous wild animals they do so in conditions which present no risk to the public and which maintain the welfare of the animals. A licence is required for any animal species listed on a schedule to the Act and is issued by the relevant local authority which may impose certain conditions on it, such as, for example, putting restrictions on the movement of the animal, and requiring the licence holder to insure himself and any other person against liability for any damage or injury caused by the animal.

DARK ADAPTED Descriptive of anurans housed in the laboratory in near or complete darkness. Frogs maintained in such conditions for some hours allow experiments to be undertaken to ascertain the effect of lack of light on skin pigmentation.

See LIGHT ADAPTED.

DAUDIN, Francois Marie, (1774-1804), French herpetologist who collaborated with Lacépède and Buffon on the *Histoire naturelle des reptiles* (1802).

DAZZLE EFFECT In snakes, the result of a combination of factors bringing about confusion and uncertainty in a predator. Such factors include cryptic, disruptive or banded coloration, skin iridescence, and swift movement through vegetation or over rough terrain.

DEATH FEINT ;
LETISIMULATION;
THANATOSIS A defensive reflex action in many amphibians and reptiles in which a position similar to that of a dead specimen is adopted. The action is displayed very well in the hognosed snake (genus *Heterodon*) which, when excited or disturbed, rolls onto its back and remains motionless with mouth agape and tongue protruding, recovering its normal posture once the disturbance has ceased. In some species the feint is evoked by touch whilst in others it is assumed without any external stimuli.

DEBRIDEMENT The elimination of necrotic tissue.

DECALCIFICATION *See* METABOLIC BONE DISORDER.

DE-CHLORINATE The action of removing chlorine from tap water by allowing the water to stand for 24 hours, or by adding a commercially available product, to dispel the chlorine naturally.

DEFENCE-FIGHT REACTION A method of intimidation in some toads, directed towards a predator, in which the limbs are extended, raising the inflated body off the ground, and the head thrust forward in the direction of the threat.

DEFENCE REACTION When most anurans are provoked or suddenly disturbed a defence reaction is evoked in which the lungs are filled with air, inflating the body, and the head bowed low.

DEFIBRINATION The process of removal or destruction of FIBRIN.

DEFRA Acronym denoting the Department of Environment Food and Rural Affairs. Based in the United Kingdom DEFRA's aim is "to protect and improve the environment and to integrate the environment with other policies across Government and in international forums." The health and welfare of animals is central to DEFRA's work of protecting and improving livestock and controlling and eradicating disease.

DEGENERATE In reference to parts of the body, or phases in the life cycle of an creature, that have become significantly reduced in size, or have gone completely, in the course of evolution.

DEGLUTITION The act of swallowing.

DEIMATIC BEHAVIOUR Intimidating or bluffing behaviour which acts as a deterrent to potential predators. In snakes it may precede a strike such as tail-vibrating in rattlesnakes and hood-spreading in cobras (intimidation) for example, and in certain toads it may increase their size by the inflation of the lungs (bluff).

DELAYED DEATH In ENVENOMATION tests, a death that ensues between twelve and forty eight hours after administration. Typically resulting from the venom's NEUROTOXIC elements. *Compare* EARLY DEATH.

DELAYED FERTILIZATION *See* AMPHIGONIA RETARDA.

DELAYED SHOCK In reference to the symptoms and effects of certain venomous snake bites, a sudden and often fatal deterioration in the victim after some days of apparent improvement.

DEN A collective noun for a gathering of snakes.

DENDROASPIN *See* MAMBIN.

DENDROBATIDAE Arrow- or dart-poison frogs. Family of the ANURA, inhabiting Central America and tropical South America. 116 species in three genera.

DENERVATION A condition in which the nerve supply is interrupted or disconnected.

DENTARY One of the major bones of the lower jaw. The dentary bone is found in all reptiles and amphibians and, in the snakes, lizards and crocodilians, bears the teeth (lacking in anurans and chelonians) of the lower jaw.

DENTARY PSEUDO-TEETH Tooth-like structures at the tip of the lower jaw in some anurans. In the tusked frogs (genus *Adelotus*), these

structures form two canine-like projections (tusks) on the lower jaws of the males.

DENTATE Serrated or toothed.

DENTICLE A small tooth or tooth like protuberance.

DENTICULATE Very finely toothed; anything having tooth- like or saw-like projections.

DENTITION The number, type, arrangement and physiology of teeth in a particular species. In reptiles and amphibians, the arrangement of teeth may be ACRODONT, PLEURODONT or THECODONT and, in snakes, may be AGLYPHIC, OPISTHOGLYPHIC, PROTEROGLYPHIC, SOLENOGLYPHIC etc.

DEPOSITION; PARTURITION The act of giving birth or laying eggs.

DEPOSITION SITE The site chosen by a female reptile or amphibian in which to lay eggs or give birth to young.

DEPRESSED In reference to body shape, flattened from top to bottom providing a greater width than depth. *Compare* COMPRESSED.

DERMA *See* DERMIS.

DERMAL LAYER The layer of skin beneath the EPIDERMIS that contains the blood vessels, nerve endings etc. of the skin.

DERMAL PLATE Any one of the LAMINA on a chelonian shell.

DERMAL POCKET *See* ACARIDOMATIUM

DERMAL TUMOUR *See* POX.

DERMATEMYDIDAE Central American river turtle. Family of the

CHELONIA, suborder Cryptodira, inhabiting Central America from Mexico to Honduras. MONOTYPIC.

DERMATITIS In the captive husbandry of reptiles, a condition in which the skin becomes inflamed and which is usually caused by keeping animals in incorrect conditions, e.g., substrate too damp, humidity too high etc. In NECROTIC dermatitis the skin, particularly that of the belly (VENTER), blisters and later sloughs away. *See* THERMAL BURN.

DERMIS The deep inner layer of the skin beneath the EPIDERMIS, containing connective tissue, blood vessels and fat.

DERMOCHELYIDAE Leatherback turtle. Family of the CHELONIA, suborder Cryptodira, inhabiting waters from around New Zealand to within the Arctic Circle. MONOTYPIC.

DEROTREMATOUS Descriptive of urodeles which, when they become adults, lose their gills whilst retaining the gill slits.

DESERT A region that is completely, or almost, lacking any vegetation due largely to extremely low rainfall. True deserts have a very sparse fauna, although they are inhabited by several reptiles and anurans.

DESICCATION Dehydration as a result of water loss from the body tissues and cells.

DETOGLOSSAL Having part of the border of the tongue fastened to a certain part of the lower jaw and therefore not free as in ADETOGLOSSAL salamanders.

Most amphibians are detoglossal.

DEVONIAN The geological period which began some 405 million years ago in the PALAEOZOIC era, between the SILURIAN and the CARBONIFEROUS periods, and was characterized by a tremendous variety and number of fish, most of which have become extinct. The Devonian period lasted until about 355 million years ago and saw the appearance of the first amphibians.

DEWLAP; GULAR EXPANSION; GULAR FAN A pendulous fold or flap of skin under the throat of some lizards and usually referring to the often extensible dewlap occurring in members of the IGUANIDAE which can, in some genera, be raised or lowered by the action of the HYOID BONE (well-developed in *Anolis*) during various behavioural displays.

DIAD; DYAD Arranged in groups of two, as in the arrangement of rings or bands on many coral snake mimics (*Lampropeltis* sp.). *See* TRICOLOUR DIAD.

DIAGNOSTIC CHARACTER Any one of the primary features differentiating a TAXON from other comparable or closely related taxa.

DIAGONAL TYPE In snakes, a SCALE ROW arrangement.

DIAPAUSE A temporary period of suspended development and growth often accompanied by reduced metabolism and correlated with seasonal change, e.g., a drop or rise in temperature, or an increase or decrease in rainfall.

DIAPSID Descriptive of a skull in which there are two TEMPORAL openings. Found in all modern reptiles except the CHELONIA.

DIAPSIDA The now redundant name formerly given to a subclass of reptiles the skulls of which are of the DIAPSID form. It has since been found that the more primitive forms and not unmistakably interrelated to the more advanced forms and consequently the diapsids are presently divided into two subclasses: the more primitive LEPIDOSAURIA (scaly reptiles) and the more advanced ARCHOSAURIA (ruling reptiles)

DIASTEMA An empty, toothless, gap in a row of teeth on the same bone.

DIBAMIDAE Blind lizards. Family of the SQUAMATA, suborder SAURIA, inhabiting north-eastern Mexico (one species) and South-East Asia, from Indo-China and the Malayan peninsula, Indonesia, and the Philippines to New Guinea. Some 10 species in two genera.

DICE Acronym denoting the Durrell Institute of Conservation and Ecology. DICE is committed to carrying out the research required to conserve biodiversity and the functioning ecosystems upon which animals, and people, depend. It was launched in 1989 as Britain's first research and postgraduate training centre in conservation science and was named in honour of the naturalist, writer and traveller, Gerald Durrell.

DICEPHALOUS Possessing two heads. *See* BICEPHALOUS.

DICHROMATIC A species having two different colour varieties that are independent of both age and sex.

DIDACTYL Having two digits, fingers or toes, on a foot, as in the South African lizard *Chamaesaura anguina*, for example.

DIGGER The enlarged TUBERCLE on the hind foot of toads of the PELOBATIDAE; the SPADE of spadefoot toads.

DIGIT A finger or toe. Characteristically, there are five digits in the basic limb structure of terrestrial vertebrates, each made up of a series of small bones (phalanges). There are however, some modifications to this plan and in many species the number of digits is reduced. Some lizards, for example, may only have two or three digits on each foot, and crocodilians have five on their forefeet and only four on the hind feet.

DIGITAL Any scale on the digits of both the fore- and hind feet, situated between the palm or sole and the claw. To show the relevant position of a scale on a digit, the term is supplemented with a prefix such as 'pre-' or 'post-' etc.

DIGITAL EXPANSION In lizards of the GEKKONIDAE, that section of the toes, and fingers, that is more or less clearly larger or wider.

DIGITAL FRINGE The enlarged, free-edged scales along the lateral margins of the longer digits of certain desert-dwelling lizards, giving a tooth-like aspect to the toe border.

DIGITAL GROOVE In certain anuran species, a deep and

conspicuous groove positioned around the horizontal edge of the PAD on the DIGITS.

DIGITAL PAD Either of the paired fleshy structures located on the upper surface of the tips of the digits in certain dendrobatid and ranid frogs and bufonid toads.

DIGITAL SCUTE Either one of the paired pads located on the upper surface of the tips of the digits in particular species of dendrobatid, bufonid and ranid anurans.

DIGITAL TREMOR The extremely rapid toe vibration which occurs in toads (*Bufo*) when watching or chasing a prey item. Male toads also vibrate the toe of their hind foot against that of the female's during AMPLEXUS.

DIGITAL WEB The membrane connecting the digits, as in the hind feet of crocodilians and anurans and in the fore- and hind feet of marine and freshwater turtles etc.

DILUTE ALBINO *See* LEUCISTIC.

DIMORPHISM *See* SEXUAL DIMORPHISM.

DINOSAUR Any of the large, extinct terrestrial reptiles that were abundant during the MESOZOIC era and which were the dominant land animals for 140 million years. Although the name means literally 'terrible lizard', dinosaurs were not lizards but DIAPSID reptiles, the closest living relatives of which are the crocodilians and birds.

DIPLOPIC In reference to the effects of certain VENOMOUS snake bites the perception of double images;

having double vision.

DIPHYCERCAL TAIL The type of tail occurring in the larvae of such amphibians as frogs and toads, in which the vertebral column extends to the tip with symmetrical caudal fins above and below it.

DIRECT DEVELOPMENT The absence of a larval stage, occurring in certain amphibians, which eliminates the many dangers of an aquatic development. Found in various frog and salamander genera throughout the world, the eggs hatch into tiny replicas of their parents, each with adult form and features.

DISC The circular, flat and enlarged adhesion surface on the tips of the digits in many arboreal lizards and amphibians, enabling them to grip smooth vertical surfaces.

DISCHARGE ORIFICE The opening near the tip of the fang of a venomous snake through which the venom from the VENOM DUCT is injected into the tissues of a bitten animal.

DISCOGLOSSIDAE Midwife toads, fire-bellied toads and related forms. Family of the ANURA, inhabiting the PALAEARCTIC, the Philippines and Borneo. There are 11 species in four genera.

DISINTEGRIN A protein component found in the venom of vipers which, although structurally unrelated, is similar to MAMBIN in preventing platelet accumulation in the blood.

DISPERSAL The distribution, for various reasons (e.g., territorial, climatic), of individual members, pairs or groups of species through a given area. Dispersal in many reptiles and amphibians occurs at the end of the breeding season when territory is forsaken and they move outwards from the breeding place either in a clearly defined direction or at random.

DISPERSAL BARRIER; ECOLOGICAL BARRIER Any area of unfavourable habitat that separates two areas of favourable habitat. Such an area could, for example, be an ocean separating two land masses, or a motorway separating two meadows

DISPLAY The obvious, showy and usually ritualized behaviour pattern, seen in courtship and territorial defence etc., that can affect the behaviour of other animals at which the display is directed. Display in reptiles is often enhanced by the aid of various body structures: the hood in cobras (*Naja*); the dewlap in iguanas (*Iguana*); the tongue in blue-tongued skinks (*Tiliqua*).

DISRUPTIVE COLORATION Colours which confuse the eye by disrupting the shape or outline of an animal against its surroundings; camouflage; CRYPTIC COLORATION. An example is well displayed by the gaboon viper *Bitis gabonica*, the striking colours of which cause it to 'disappear' when lying amongst leaf litter on the forest floor.

DISRUPTIVE OUTLINE A particular shape which confuses the eye by disrupting the shape or outline of an animal when seen against its surroundings. An example is well displayed by the Asian horned frog (*Megophrys*), the shape of which helps to conceal it as it sits amongst the leaves of the forest floor.

DISSEMINATED INTRA-VASCULAR COAGULOPATHY In reference to the effects of certain venomous snake bites, an uncharacteristic coagulating condition which exhausts stores of blood clotting factors and platelets and results in incoagulable blood thereby giving rise to the possibility of severe bleeding.

DISTAL Describing the part of a limb, organ or appendage etc., that is located farthest from the centre, median line, or point of attachment or origin. *Compare* PROXIMAL.

DISTENSION Swelling, inflation or expansion, as occurs in the throats of male anurans when making their CALL, or in the throats of many snakes when swallowing large food items.

DISTRIBUTION The location, or occurrence, of a species, also known as its 'range', and usually referring to the pattern formed when plotted on a map.

DISTRIBUTIONAL BARRIER Referring to an ecological or geographical obstacle or barrier that, for particular animal groups, cannot be overcome, thereby preventing their DISPERSAL and the subsequent extension of their range. Deserts, mountain ranges, oceans and large lakes are distributional barriers.

DITMARS, Raymond Lee, (1876-1942), American herpetologist. Worked as Curator of Mammals and Reptiles at New York's Bronx Zoo. Undertook many expeditions and successfully bred many of the animals he collected. One of the first to popularise herpetology in North America, his publications include *Snakes of the world* (1931), *Snake-hunters' holiday* (1935, with W. Bridges), *The reptiles of North America* (1936), *Field book of North American snakes* (1939).

DIURNAL Active during the daylight hours.

DIURNAL RHYTHM Any activity cycle in an organism that is related to biological processes which occur regularly at 24-hourly intervals. *See* BIOLOGICAL CLOCK; CIRCADIAN RHYTHM.

DIVERGENT EVOLUTION The gradual development, through evolution, of a number of different varieties or species of animals (or plants) from a common ancestor, each adapting to new habitats and different sources of food. *See* ADAPTIVE RADIATION.

DIVERSITY *See* SPECIES DIVERSITY.

DIVERTICULUM (pl.) **DIVERTICULA** An elongated pouch with only one opening; diverticulate lungs have many blind branches, as opposed to a single large sac.

DoE Acronym denoting the Department of the Environment (United Kingdom).

DOME The term given to the attitude adopted by female fence lizards (genus *Sceloporus*), which is usually stimulated by the appearance of males and in which the body is raised off the ground on stiffened legs, bending the back into a hump, or 'dome', frequently supplemented with short, jerky jumps.

DOMINANCE The community hierarchy (peck order) between and within species and referring to the social standing of an individual within a group. A dominant lizard, for example, is one that is allowed priority in access to food, mates etc., by other members of its species as a result of its size, strength or success in previous aggressive encounters.

DORSAL Relating to the upper surfaces; the back or spinal part of the body. **1.** Any one of the enlarged scales on the upper surface of the body of crocodilians. **2.** Any one of the scales on the backs of lizards. **3.** Any one of the scales on the upper surface of the body of most snakes (i.e., all scales except the VENTRALS).

DORSOLATERAL Referring to markings, colours, scales, appendages etc., which are not directly down the centre of the back, nor on the side of the body, but more or less between the two.

DORSOLATERAL FOLD In certain anurans, a longitudinal, glandular ridge located between the centre of the back and the side of the body.

DORSUM Of, or relating to, the upper surface of an animal; e.g., the dorsum of the neck is the NAPE.

DOUBLE-CLUTCHING Said of many reptiles that produce two clutches of eggs or young in the same season (i.e., within one year).

DRACO The flying lizards, genus of the AGAMIDAE, inhabiting the Indo-Australian Archipeligo, South-East Asia and the Philippines, in some 20 species. Identified by the characteristic skin folds (WINGS) on each side of the body which, when spread, form gliding membranes enabling the lizards to glide for distances of over 20 metres between trees.

DRAW To extract the fangs from a venomous snake. Snake-charmers in many parts of the world sometimes render their animals temporarily harmless by drawing their fangs prior to a display.

DTH Abbreviation for DAYTIME HIGH.

DTL Abbreviation for DAYTIME LOW.

DUCT Any bodily tube, channel or passage through which a fluid, especially a secretion or excretion, moves, e.g., VENOM DUCT.

DUD The term for an egg which fails to be laid and in which embryonic development has ceased for one reason or another, causing it to shrivel and harden within the OVIDUCT.

DUGITE *Pseudonaja affinis*, an extremely dangerous and aggressive snake of the family ELAPIDAE, inhabiting the south-western corner of Western Australia and extending eastwards along a narrow strip of coast to the South Australian border.

DUMÉRIL, André Marie Constant, (1764-1860), French herpetologist and anatomist. Published the first comprehensive and systematic work on the reptiles and amphibians in the form of the impressive ten-volume *Erpetologie generale ou Histoire naturelle complete des reptiles*, in co-operation with Gabriel Bibron and, in part, his son, Auguste H. A. Duméril.

DUMÉRIL, Auguste H. A., (1812-1870), son of the distinguished herpetologist, André M.C. Duméril. Compiled, in collaboration with Brocchi, Mocquard and Bocourt, the 17-volume systematic work entitled *Mission scientifique au Mexique et dans l'Amerique Centrale: Etudes sur les reptiles et les batraciens* (1870-1909), which became the foundation for Latin-American herpetology thereafter.

DUNN, Emmet Reid, (1894-1956), American herpetologist. Publications include *Salamanders of the family Plethodontidae* (1926), *American frogs of the family Pipidae* (1948).

DUPLICATION Referring to the copying or multiplication of different parts of the body as a result of genetic interferences. Duplication in several forms has occurred in amphibians, urodeles in particular, examples of which include an individual with two heads, two individuals with one head between them, and specimens with extra limbs.

DUVERNOY'S GLAND The venom-producing gland of rear-fanged colubrid snakes, usually situated between the eye and the angle of the mouth; named after the French anatomist, D. M. Duvernoy.

DYAD An alternative spelling for DIAD.

DYSECDYSIS The condition in which a reptile has difficulty sloughing its skin.

DYSFUNCTION Impairment of normal function.

DYSPNOEA In reference to the effects of certain venomous snake bites, a shortness of breath, usually in response to low blood oxygen levels.

DYSTOCIA Difficulty in giving birth to young or laying eggs. In the case of the latter, the animal is said to be suffering from EGG-BINDING.

E

EARDRUM *See* TYMPANUM.

EAR FLAP; OPERCULAR FLAP
Either one of the large upper or small lower skin folds situated at the outer opening of the auditory canal of crocodilians. When the reptile is submerged the flaps, or ear lids, are kept tightly shut, preventing water from entering the hearing organs. On land, the upper flap is lifted revealing a narrow transverse opening which admits sound.

EAR LID *See* EAR FLAP.

EARLY DEATH In ENVENOMATION tests, a death that ensues within eight hours after administration. Typically resulting from the venom's HAEMOTOXIC elements. *Compare* DELAYED DEATH.

EARTHWORM MOVEMENT
Alternative term for the CONCERTINA MOVEMENT of snakes.

ECARIN A prothrombin-converting enzyme isolated from the venom of the saw-scaled viper, *Echis carinatus*.

ECARINATE Descriptive of scales or laminae which do not bear a keel and are therefore entirely smooth.

ECCHYMOSIS In reference to the effects of certain venomous snake bites, a bruise-like discoloration of the skin resulting from small dispersed blood clots.

ECDYSIS The act of periodically shedding the outer layer of dead, keratinous skin, the STRATUM CORNEUM, to permit further growth in both reptiles and amphibians.

ECDYSIS CYCLE *See* SLOUGH CYCLE.

ECHINULATE Having a covering of small points or spines.

ECOLOGICAL BARRIER *See* DISPERSAL BARRIER.

ECOLOGICAL FACTOR
Referring to any one of a number of environmental elements important in the lives of reptiles and amphibians, such as competitors, enemies and parasites, as well as food, water, humidity, temperature etc.

ECOLOGICAL NICHE *See* NICHE.

ECOLOGICAL SYSTEM *See* ECOSYSTEM.

ECOLOGY The study of the interactions of animals and plants in relation to their environment and to each other.

ECOSYSTEM; ECOLOGICAL SYSTEM The system involving the entire environment and the interactions between the living community of plants and animals, directly or indirectly, with the non-living entities and the elements

ECOTONE The transition zone between two ecological communities. A woodland edge, for example, is an ecotone separating an open heathland habitat from a forest habitat, along with the particular reptile and amphibian communities which they support.

ECR Abbreviation for EXCITEMENT COLOUR REACTION.

ECTO- A prefix meaning outside; outer; external.

ECTOGLYPH 1. A venomous snake possessing fangs which are located at the front of the mouth, with channelled grooves down which venom flows into the tissues of the victim.
2. A snake possessing teeth which are located at the rear of the mouth, with channelled grooves.

ECTOPARASITE Any parasite, such as a tick or mite (order Acarina), which attaches itself to the outer part of the body of its host in order to extract nourishment from blood vessels.

ECTOPARASITOSIS Symptoms of disease resulting from infestation with an ECTOPARASITE, which can be a VECTOR of a number of diseases affecting reptiles.

ECTOPIC Out of position; structures or parts that are displaced. Used in herpetology for eggs that have undergone the early development stages in a different place than the OVIDUCT.

ECTOTHERM; COLD-BLOODED; POIKILOTHERM An animal having a constant self-regulated body temperature and dependent entirely upon outside heat sources.

EDAPHIC Of, or related to, the ground. The chemical and physical composition of the soil, together with other factors, such as its temperature and water content, play a significant role in the lives of reptiles and amphibians, from egg deposition in or upon the soil, to hibernation below ground.

EDEMA *See* OEDEMA.

EDENTULOUS Without teeth, referring in particular to the condition of having lost teeth which were once present.

EEC Acronym denoting the European Economic Community.

EFT An antiquated, dialectal, or local name for any of the various small semi-aquatic urodele amphibians, e.g., the European smooth newt *Triturus vulgaris*, and, in many regions, particularly their immature larvae.

EGG-BINDING: DYSTOCIA A condition occurring in reptiles in which eggs fail to be laid and are retained within the oviduct. Causes include physical abnormality or damage and/or incorrect environmental conditions.

EGG CALLUS *See* EGG TOOTH.

EGG CAPSULE *See* JELLY ENVELOPE.

EGG ENVELOPE The jelly-like membrane that surrounds the egg in amphibians.

EGG MASS The whole amount of eggs deposited, in a clump or string, by a single amphibian female. *See* SPAWN.

EGG RETENTION *See* EGG-BINDING.

EGG SAC The protective membrane which adheres to the eggs of certain urodeles, e.g., *Ambystoma, Salamandrella,* as they pass through the oviduct.

EGG TOOTH; CARBUNCLE; EGG CALLUS The forward-pointing, foetal tooth located on the end of the premaxillary bone in OVIPAROUS snakes and lizards, used in breaking through the egg shell at birth. The egg tooth is deciduous and falls off shortly after hatching.

ELAPID Any snake of the mainly tropical family ELAPIDAE possessing fixed venom-conducting fangs at the front of the upper jaw. The family includes the mambas, cobras and coral snakes.

ELAPIDAE Cobras and their allies. Family of the SQUAMATA, suborder Caenophidia, inhabiting most regions of the world except the HOLARCTIC. Over 160 species in over 50 genera.

ELECTRO-EJACULATION The collection of sperm from a male reptile by means of a mild electric current, and for use in ARTIFICIAL INSEMINATION.

ELECTROPHORETIC PATTERN A configuration created by the application of electric fields to a substance in order to generate dispersion.

ELEVATION The term for a medical procedure in which a swollen limb is raised above the level of the heart in order to encourage the swelling's reduction.

ELISA Acronym denoting Enzyme Linked ImmunoSorbant Assay; a method of determining which snake species was responsible for a particular snakebite by comparison of the victim's blood serum with venoms from known snake specimens.

ELKAN, **Edward,** (1895-1983), German physician and zoologist who pioneered lower vertebrate pathology and gained an international reputation for his work on lower vertebrate disease. Published numerous papers and chapters, and co-authored with Professor Reichenbach-Klinke the classic *The principal diseases of lower vertebrates* (1965). In 1983 the Edward Elkan Memorial Fund was established and his collection of numerous photographic slides, specimens, drawings, books and papers have formed the Edward Elkan Collection of Lower Vertebrate Pathology.

ELLIPTICAL Of, or having the form of, an ellipse (a closed curve in the form of a flattened circle), and used to describe the VERTICAL pupils in many reptiles and amphibians.

EMACIATION The condition seen in some captive, or recently imported, reptiles and amphibians indicative of inadequate metabolism or hastened energy loss and dehydration, as a result of incorrect husbandry, injury or shock, or a combination of these factors.

EMARGINATE Pertaining to a border that is notched or scalloped in outline and used, in herpetology, to describe an arrangement where the edge of a scale cuts into the otherwise smooth border of a larger one.

EMBRACE *See* AMPLEXUS.

EMBRYONIC DEVELOPMENT
The progress of the early stages
(embryo) of an animal, beginning
with fertilization of the egg cell and
ending at birth or hatching.

EMERALD NETWORK A network
of areas of special conservation
interest (ASCIs) which is to be
established in the territory of the
contracting Parties and observer
States to the BERN CONVENTION
including, among others, central and
east European countries and the
European Union Member States. For
EU Member States, Emerald
Network sites are those of the
NATURA 2000 network.

EMYDIDAE Freshwater turtles.
Family of the CHELONIA, suborder
Cryptodira, inhabiting all continents
except Australia. 136 species in
some 30 genera.

ENCAPSULATED Enclosed in a
capsule, or membrane, such as the
embryo of an amphibian.

END-BODY *See* MATRIX.

ENDEMIC Descriptive of an animal
(or plant) that is restricted to a
biogeographic region, or part of it
and, in some cases, originating there.

ENDO- A prefix meaning internal;
inner; within.

ENDOBIOTIC Developing within a
living organism. Descriptive of an
ENDOPARASITE.

ENDOCRINE SYSTEM *See*
ENDOCRINE GLAND.

ENDOCRINE GLAND Any one of
the glands in an animal that produce
hormones and secrete them directly
into the bloodstream, including the
PITUITARY GLAND, TESTIS and

OVARY etc., and collectively form
the 'endocrine system'.

ENDOGENOUS Internally; coming
from, or developing from, within the
animal. Growth rhythms which are
not regulated by various
environmental stimuli are called
'endogenous rhythms'.

ENDOGLYPH 1. A venomous snake
possessing fangs at the front of the
mouth, with closed grooves through
which venom flows into the tissues
of the victim. **2**. A snake possessing
teeth, located at the front of the
mouth, with closed grooves.

ENDOLYMPHATIC SAC Either
one of the pair of sacs located one on
each side of the neck and containing
calcium carbonate, in certain
members of the subfamily
gekkoninae (day geckos).
See CALCIUM CYCLE.

ENDOPARASITE Any parasite,
e.g., a FLUKE or TAPEWORM, that
extracts nourishment from the
internal parts of the body of its host.

ENDOPARASITOSIS Symptoms of
disease resulting from an infestation
with an ENDOPARASITE which
can affect the internal organs of
reptiles and amphibians.

ENDOSKELETON The internal
skeleton of an animal, e.g., the bony
or cartilaginous skeleton of
vertebrates, which supports the body,
protects the vital organs, and
provides a system of levers on which
muscles can act and produce
movement. The endoskeleton
consists of a skull, vertebral column,
pectoral and pelvic girdles, ribs, and
limb elements, and permits growth in

the animal.

ENDOTHELIUM A flat cellular layer lining the inner walls of blood vessels.

ENDOTHERM An animal that maintains a constant body temperature via inherent, internal heat sources; a warm-blooded animal or HOMOIOTHERM.

ENDYSIS A term used occasionally to describe the formation of a new epidermal layer. *Compare* ECDYSIS.

ENEMY REACTION A particular response, in defensive or flight behaviour, evoked by stimuli from a predator or rival, or other threat to the safety of an animal. Examples of enemy reaction among reptiles and amphibians include the DEATH FEINT, hood-spreading in cobras, and the assuming of the BALL POSITION in many snakes, among others.

ENTAMOEBIASIS *See* AMOEBIASIS.

ENTERIC Of, or in, the intestinal cavity. Many parasites are enteric in reptiles and amphibians.

ENTEROLITH A stone or concretion found in the intestine. *See* BEZOAR STONE.

ENTOPLASTRON The only single, centrally located bone in the PLASTRON of most chelonians. Diamond-shaped, it occurs between, and is surrounded by, the paired EPIPLASTRON and HYOPLASTRON bones.

ENVENOMATION; VENENATION A condition resulting from having VENOM injected into the tissues of the body, as in a case of a venomous SNAKE BITE.

ENVIRONMENT The entire surroundings and range of conditions in which an animal or plant lives including biological, chemical and physical factors such as light, temperature, humidity and the availability of water, food etc., which more or less influence its behaviour and development.

ENZOOTIC Pertaining to diseases which affect animals within a limited region. *Compare* EPIZOOTIC.

ENZYME Any one of a group of complex proteins that are formed by living cells and bring about or strengthen specific biochemical reactions.

ENZYME LINKED IMMUNE SORBANT ASSAY *See* ELISA.

EOCENE The second oldest geological epoch of the TERTIARY period. This lasted for some 20 million years, from 55 million to 38 million years ago, and was characterized by the predominance of early hoofed mammals (the ungulates), although many others, including early horses, rodents, carnivores and whales, were also making an appearance.

EPANODONT Having teeth on the lower jaw only, the upper jaw bones being EDENTULOUS.

EPI- A prefix meaning above; upon; in addition to.

EPICORACOID CARTILAGE A rear component of the CORACOID cartilage bones in the PECTORAL GIRDLE.

EPIDEMIOLOGY The branch of science that deals with the incidence, transmission, distribution and control of epidemic diseases in a population.

EPIDERMIS The outermost protective layer of cells of the body of an animal. In vertebrates it consists of several layers of cells and forms the external layer of skin, protecting the underlying tissues and ensuring that the body is waterproof. As it is worn away at the surface it is continually replaced, in the MALPIGHIAN LAYER immediately beneath, by the growth of new cells (ECDYSIS and ENDYSIS in reptiles and amphibians).

EPIDIDYMIS The long narrow coiled tube, derived from part of the embryonic kidney, attached to the surface of the TESTIS in male reptiles, which serves as a temporary storage organ for spermatozoa until their release to the exterior, via the VAS DEFERENS, during mating.

EPIGAMIC The feature of a reptile or amphibian that entices or excites members of the opposite sex during courtship. An example is the colourful crested tail of the male great crested newt *Triturus cristatus*.

EPIPHYSIAL EYE *See* PINEAL COMPLEX.

EPIPHYTE A plant, such as a moss, fern or orchid, which grows on another plant, usually the branch or trunk of a tree, but is not parasitic on it.

EPIPLASTRON (pl.)
EPIPLASTRA Either one of the foremost pair of bones of the PLASTRON of a chelonian. (In the plastron of soft-shelled turtles (TRIONYCHIDAE) the bone is not paired and is generally referred to as the ENTOPLASTRON.

EPIPODIAL Referring to any one of the bones of the lower forelimb (forearm) or the lower hind limb (lower leg), i.e., the RADIUS and ULNA (forelimb; FIBULA and TIBIA (hind limb).

EPIPTERYGOID BONE Either one of two vertical struts that support the bones of the palate in lizards.

EPIPYGAL *See* SUPRAPYGAL.

EPISTAXIS In reference to the effects of certain venomous snake bites, a haemorrhage from the nose; nosebleed.

EPIZOIC Growing or living on the exterior of a living animal. Ticks and mites, of the order Acarina, are epizoites, living on reptiles.

EPIZOOTIC Pertaining to a disease that suddenly and temporarily affects a large number of animals. Compare ENZOOTIC.

EREMIAL Any region in which the predominant weather conditions, particularly low rainfall and adverse winds, prevent the growth of scrub and tree cover. For example, rocky deserts, heathlands, prairies etc., are eremial regions and all good habitats for many reptile species and certain amphibians.

ERYTHEMATOUS In reference to the effects of certain venomous snake bites, the leakage of blood into adjacent tissues; bruising.

ERYTHRISM An excess of reddish pigment in an individual or population compared with its normal species coloration. An erythristic individual is a colour phase in which the dominant colour is red, giving it an ABERRANT appearance.

ERYTHROCYTIC Pertaining to red blood cells.

ERYTHROPOIESIS The formation of red blood cells.

ESCUTCHEON In the males of certain gekkonids (*Eublepharis*) an area of ventral scales, directly in front of the VENT and on the under surfaces of the hind limbs, that are specialized, thick and frequently sharply contrasting in colour to the other VENTRAL scales. It is less noticeable in females.

ESR Abbreviation for EXCITEMENT SECRETION REACTION.

ESTIVATION An alternative spelling of AESTIVATION.

ETHIOPIAN REGION One of the six zoogeographical regions of the world, contrived in conformity with land-animal distribution. The Ethiopian region is made up of Africa south of the Sahara and the Malagasy sub region, which consists of Madagascar, the Mascarenes, Seychelle Islands and the southern tip of the Arabian Peninsula. Reptiles and amphibians are well represented in the region in widely distributed and diverse forms and almost all of the CHAMAELEONIDAE occur in it.

ETHOGRAM A detailed record and description of every aspect of behaviour within a particular species, observed in the field or in captivity, which when published can help greatly to extend present knowledge of an animal's behavioural characteristics.

ETHOLOGY The study of the behaviour of reptiles and amphibians (and other animal species) in their natural environment.

EUBLEPHARID Referring to any one of a group of gekkonid lizards which possess prominent movable eyelids, digits which have claws but no LAMELLAE, and a slow stalking gait. Together they form the Eublepharinae, a subfamily of the GEKKONIDAE.

EUPLEURODONT The term given to PLEURODONT reptiles in which new teeth lie in a vertical row at the base of functioning teeth, moving into empty sockets as and when others are lost (a process termed 'vertical replacement'). Compare SUBPLEURODONT.

EURY- A prefix meaning wide; broad.

EURYAPSID Describing a skull that has a single upper temporal opening. Found in some fossil saurians of the extinct subclass Euryapsida.

EURYECIOUS Having a wide variety of habitats

EURYPHAGIC Said of any animal able to devour a wide variety of food.

EURYTHERMAL Said of any animal able to endure a wide variation in temperatures. For example, the leopard gecko *Eublepharis macularius* thrives in temperatures which may be as high as 38°C during the day and as low as

70

15°C at night. *Compare* STENOTHERMAL.

EURYTOPIC Able to tolerate a wide range of a number of factors.

EUSTACHIAN TUBE The tube connecting the middle ear to the back of the throat in vertebrates. Normally closed, it opens during the acts of swallowing and yawning to allow air into the middle ear, thus maintaining atmospheric pressure on each side of the eardrum (tympanum). The tube was named after Bartolomeo Eustachio, the Italian anatomist.

EUTHANASIA The act of putting to death in a humane manner. Reptiles and amphibians are euthanized in order to relieve suffering from an incurable disease or mortal wound. Also called mercy-killing.

EUTROPHICATION The nutrient improvement of bodies of water caused by organic enhancement. Although a natural process, fast eutrophication can exhaust oxygen in the water and may result in the death of aquatic organisms.

EURYOECIOUS Descriptive of any species able to withstand a wide variety of conditions or has no specific requirements, e.g., a eurythermal frog species may accept a broad range of water temperatures. *Compare* STENOECIOUS.

EVAGINATION The act of turning inside-out.

EVOLUTION A gradual, long-term change in a population's genetic structure and other characteristics, e.g., anatomy, physiology, behaviour etc., over successive generations that allows its members to effectively adapt to environmental conditions and to better utilize food sources.

EVOLUTIONARY LINEAGE The processional line of descent of a TAXON from its ancestral taxon. Such a lineage stretches back, through the different taxonomic levels, from SPECIES to GENUS to FAMILY to ORDER etc.

EVOLUTIONARY SPECIES A concept developed by those who use comparison of DNA sequences to measure relatedness between species and identify their common ancestors. The use of DNA sequences to reveal species' evolutionary history has already other thrown some long-established classification systems and has, for example, led to a complete revolution in ideas about relationships between flowering plant families. *See* BIOLOGICAL SPECIES; TAXONOMIC SPECIES.

EXANAL Referring to any of the scales located around the lateral margins of the anus in the sunbeam snake *Xenopeltis unicolor*

EXCISION The surgical removal of an organ, structure or part.

EXCITEMENT COLOUR REACTION An immediate or swift alteration in pigmentation of the skin as a direct response to various external stimuli.

EXCITEMENT SECRETION REACTION In certain amphibians, the comparatively swift and liberal production of skin mucus as a direct response to various external stimuli.

EXCLUSION In VIVIPAROUS members of the SALAMAN-DRIDAE, the process of giving birth.

EXCRESCENCE Any natural projection or protuberance and, in particular, an epidermal outgrowth from an organ or part of the body. Frequently used, in herpetology, for many of the various structures occurring on the bodies and limbs of reptiles and amphibians, e.g., the anuran NUPTIAL PAD.

EXCRETION The removal of harmful waste products, created by metabolic activities, from the blood by the kidneys in the form of urine, which is evacuated with the faeces through the CLOACA.

EXCRETORY ORGAN Any one of a system of specialized organs, the function of which is to eliminate metabolic waste products, nitrogenous compounds, carbon dioxide, water, salts etc., from the body. Examples of excretory organs in vertebrates are: the kidneys (for nitrogenous compounds, e.g., UREA and water), the lungs (for water and carbon dioxide), the urinary ducts, and the urinary bladder. In chelonians and anurans the urinary bladder has a secondary function, acting as a water-storage organ. In all snakes, amphisbaenids, varanids and crocodilians, it is entirely absent.

EXFOLIATION The act of moulting; the peeling off of skin in layers, flakes, scales or plates.

EXOGENOUS Externally; developing or originating outside an animal. External factors that, in some way, influence an animal, e.g., temperature, humidity and light, are exogenous.

EXOPTHALMOS; EXOPTHALMUS; EXOPTHALMIA The abnormal protrusion of the eyeball. Can occur in both reptiles and amphibians and is usually related to a disease of the thyroid gland.

EXOSKELETON The protective or supporting structure which covers the outside of the body of some vertebrates and provides protection for internal organs and attachments for muscles. In reptiles the exoskeleton manifests itself in the form of scales and laminae and, in chelonians, as the shell.

EXOTIC Not INDIGENOUS to a particular area; non-native. Many exotic species including amphibians and reptiles are now found living in the wild in various parts of the world to which they are alien because of introduction, either accidental or intentional, by man.

EXOTIC SPECIES *See* EXOTIC.

EXOTOXIN Any TOXIN secreted by an animal.

EXOVATION A term used to describe the evacuation of the egg; the act of hatching.

EXSANGUINATION Severing the blood vessels of an animal and allowing it to bleed to death. A method of humane killing of reptiles and amphibians (which are preferably already unconscious); a form of EUTHANASIA.

EXTANT Pertaining to a taxon some of whose members are existing at the present time. *Compare* EXTINCT.

EXTENSILE Capable of being stretched or extended.

EXTERNAL EAR OPENING The part of the ear, consisting of the auricle and the auditory canal, that leads to the outside of an animal's head.

EXTERNAL FERTILIZATION The vital process of sexual reproduction occurring outside of the female's body and usually an adaptation to life in an aquatic environment. External FERTILIZATION occurs in most ANURANS.
Contrast INTERNAL FERTILIZATION.

EXTERNAL GILL In amphibian larvae, the structures of respiration, located on the side of the neck which, at METAMORPHOSIS, are reabsorbed in most groups although in some, such as the mud-puppies, *Necturus* etc., they remain permanently throughout adulthood.

EXTERNAL NARES *See* NARIS.

EXTINCT Pertaining to a taxon none of whose members are existing at the present time. With so many threats to the environment many reptile and amphibian species are facing worldwide EXTINCTION.

EXTINCTION The act of eliminating a TAXON or the state of being eliminated

EXTRAVASATION In reference to the effects of certain venomous snake bites, the escape of body fluids, such as lymph or blood, from their proper vessels into surrounding tissues.

EXUDATE Any discharge, or exuded substance, such as cellular debris, from skin pores or wounds etc.

EXUVIA (pl.) **EXUVIAE; EXUVIUM** A LAMINA shed from a chelonian CARAPACE or a layer of skin shed by a snake, lizard, or amphibian during ECDYSIS.

EXUVIAL GLAND A skin gland in reptiles that releases an oily secretion between the outer epidermal layer and the next layer prior to skin shedding, serving to free it and thus assist its removal. The secretion causes the eyes of snakes to become opaque and take on a cloudy bluish hue. Such snakes are said to be MILKED UP.

EXUVIATION The act of shedding, moulting or sloughing the skin or similar external covering. In reptiles and amphibians, the exuviae sloughed during ECDYSIS consist of the scales and outer epidermal layers which covered their bodies, including, in chelonians, the laminae on their shells.

EXUVIUM *See* EXUVIA.

EYE SPOT 1. Any one of the rounded, often brightly coloured spots occurring in the skin patterns of some reptiles, especially lizards. 2. Rounded glandular areas of striking and contrasting colour, occurring in pairs, on the rumps of a number of South American leptodactylids, including *Telmatobius praebasalticus, Physalaemus nattereri* and some *Pleurodema* species. These eye-like spots, or OCELLI, are displayed in defence behaviour as an intimidation to enemies. Eye spots, in various forms, also occur in many other anurans.

EYE STRIPE A line that passes
'through' the eye of many reptiles
and amphibians as seen, for example,
in the European smooth snake
Coronella austriaca.

F

f Abbreviation for female.

F; °F Degrees Fahrenheit.

F₁; F₁ GENERATION *See* FILIAL GENERATION.

F₂; F₂ GENERATION *See* FILIAL GENERATION.

FACE-OFF POSTURE The position assumed by the males and, less frequently, females of many lizard species, in which two individuals present their flanks to one another, usually whilst facing opposite directions and often accompanied by the inflation or compression of the body. The face-off posture is used during the defence of territory or food, or during fights over females etc., and can be observed particularly well in lizards of the genus *Tiliqua*.

FACIAL PIT *See* LOREAL PIT.

FACULTATIVE Having the capability to live under more than one set of environmental conditions; capable of, but not obligated to, a specific physiological process (e.g., hibernation), or a specific mode of life (e.g., subterranean).

FAECAL ANALYSIS Examination of the waste products of reptiles and amphibians to determine the presence of disease or parasitic worms etc.

FAECES; FECES The solid or semi-solid material that is eliminated from the digestive tract through the anus, and consists of indigestible food residues, bacteria, bile, mucus and dead cells from the gut lining.

FALCATE Curved; shaped like a sickle.

FALSE GHARIAL *See* GHARIAL

FAMILY The taxonomic category used in the classification of organisms that consists of one or several similar or closely related genera, and ranking between ORDER and GENUS. The scientific name of an amphibian or reptile family always ends in 'idae', e.g., Bufonidae (toads) and Varanidae (monitor lizards), and is derived from a type genus (*Bufo* and *Varanus* in the examples above) that is characteristic of the whole family.

FAN The prominent elevated and laterally compressed helmet-like crest on the back of the head in lizards of the *Basiliscus* genus.

FANG Any of the recurved, lengthened, hollow or grooved teeth located on the MAXILLA of the upper jaw in snakes. It is usually longer than the other teeth and serves to conduct VENOM.

FANG SUCCESSION In a VENOMOUS snake the regular and sequential substitution of the functional fangs if and when they are lost by REPLACEMENT FANGS.

FASCIA (pl.) FASCIAE A sheet of fibrous connective tissue beneath the skin's surface and between muscles and groups of muscles.

FASCICULATION The uncontrolled and repeated contraction of muscles.

FASCICULIN A toxin component found in the venom of African mambas (*Dendroaspis* sp.). Although fasciculins are, comparatively, only weakly toxic requiring, for example, 50 times or more than a CURAREMIMETIC TOXIN to kill a mouse, they can cause death by paralysing the respiratory system.

FASCIOTOMY In the treatment of snake bite, an incision sometimes made to release pressure from severe swelling.

FAT BODY In amphibians and reptiles, one of a pair of organs in the abdominal cavity attached to the kidneys or near the rectum, each a mass of fatty (ADIPOSE) tissue serving as an energy store for use during hibernation or periods when food is in short supply. Also used during breeding, in the formation of sexual products, they become greatly enlarged prior to mating.

FAT CYCLE The process in which fat bodies are formed and increase in size, followed by their reduction, as the yolk formation within the eggs progresses, to small structures mainly made up of connective tissue.

FAUNA All the animal life of a given time, locality or region. The reptile and amphibian component of the fauna is termed the HERPETOFAUNA.

FAUNAL ELEMENT Used in reference to an animal species that has evolved within one of the world's six ZOOGEOGRAPHICAL REGIONS and has subsequently become a distinguishing feature of that region. The tuatara (*Sphenodon*) is, for example, a faunal element of the AUSTRALASIAN REGION.

FAUNAL REGION *See* ZOOGEOGRAPHICAL REGION.

FAUNISTICS The classification of organisms with regard to animals only. A branch of taxonomy.

FECES *See* FAECES.

FECUND Productive; fertile; capable of producing eggs.

FECUNDITY Fruitfulness; fertility; having the capability of producing eggs or young.

FEMORAL **1.** Referring to any of the scales on the upper part (thigh) on the hind limb of lizards. **2.** Either one of the second rearmost pair of plates (laminae) situated between the ABDOMINAL and ANAL plates on the PLASTRON of chelonians.

FEMORAL PORES Small, but comparatively deep, openings in the centres of certain enlarged scales on the undersides of the thighs in some lizards. The pores contain a wax-like material consisting of cellular debris which may, in breeding males, project from the scales' surface, forming a COMB.

FEMUR (pl.) **FEMORA** The main long bone forming the upper or thigh bone of the hind limb, extending from the knee to the pelvis.

FER DE LANCE A group of snakes, of the genus *Bothrops,* named after their characteristic lance-shaped heads by French settlers. Today the name fer de lance is normally associated with just one species, *Bothrops lanceolatus,* a PIT VIPER of the family VIPERIDAE, subfamily Crotalinae, found exclusively on the island of Martinique in the West Indies.

FERTILIZATION When a sperm from a male penetrates the OVUM of a female; the union of male and female GAMETES during sexual reproduction to form a ZYGOTE which will ultimately grow into a mature form of the parent organisms. In OVIPAROUS reptilian species fertilization takes place when sperm meet the OVA as they travel through the OVIDUCT, a passage which also coats the outside of the egg with calcium to produce a shell. *See* INTERNAL FERTILIZATION and EXTERNAL FERTILIZATION.

FFPS Acronym denoting the Flora and Fauna Preservation Society (United Kingdom).

FIBRIN A filamentous protein fashioned by the action of THROMBIN on FIBRINOGEN; the foundation of blood clot formation.

FIBRINOGEN A soluble protein, a globulin, in blood plasma, converted to fibrin by the action of the enzyme thrombin when blood clots.

FIBRINOGENOLYSIS The destruction of FIBRINOGEN.

FIBRINOLYSIN In reference to the effects of certain venomous snake bites, a substance that breaks down produced blood clots, thus encouraging bleeding.

FIBULA The outer and smaller of the two long bones of the lower hind limb, extending from the knee to the ankle.

FIELD IDENTIFICATION The verification of an individual specimen's classification under field conditions often with the aid of a KEY.

FIELD PLANKTON Invertebrates collected from areas of the wild, unpolluted by herbicides and insecticides, for use as food for captive reptiles.

FILIAL GENERATION Filial, derived from the Latin *filia* meaning 'daughter' and *filius* meaning 'son'. The filial generation of organisms is the first generation (F_1) of progeny resulting from the crossing of two specific parental lines, known as the P generation. The second filial generation (F_2) results from the self-crossing, or interbreeding, of the offspring from the F_1 generation. Succeeding generations of offspring (F_3, F_4 etc.) result from a sequence of controlled intercrossings with the progeny of each new F_1 generation. *See* HYBRIDIZATION; INTERGRADE; PARENTAL GENERATION.

FILARIA Any parasitic, thread-like nematode worm of the family Filariidae, a few of which are to be found in reptiles, especially certain lizards and crocodilians. Transmitted through the bites of insects, they live in the blood and tissues and spread throughout the body, heavy infestations almost always resulting in death.

FILIFORM Threadlike; like a filament.

FILTER DISPERSAL ROUTE An expression, introduced in 1940 by the American palaeontologist G. G. Simpson, to illustrate a faunal migration route across which the spread of certain animals is highly probable whereas the spread of others is respectively highly improbable. The route consequently filters out part of the fauna, but permits the rest to pass. Oceans, deserts and mountain ranges present examples of such filter dispersal routes. *See* CORRIDOR and SWEEPSTAKE DISPERSAL ROUTE, and SEASONAL MOVEMENT.

FIN Any laterally flattened, membranous organ on the body and tail of aquatic vertebrates, such as amphibian larvae, used for balance and locomotion when swimming.

FIRMISTERNAL Descriptive of the pectoral girdle in anurans, in which the EPICORACOIDS are united in the mid-line of the sternum.

FITZINGER, **Leopold**, (1802-1884), Austrian herpetologist at Vienna's Imperial Natural History Museum and author of numerous taxa. His work and publications contributed greatly to herpetological knowledge and included *Neuen Classification der Reptilien und ihren Natuerlichen Verwandschaften* (1826), and the much-admired *Bilder-Atlas* (1850).

FITZSIMONS, **Fredrick William**, (1870-1951), South African herpetologist and son of V.F.M. FitzSimons. A noted authority on South African snakes, his many publications included *Snakes of southern Africa* (1919), *Pythons and their ways* (1930), and *Snakes and treatment of snakebite* (1932).

FITZSIMONS, **Vivian Frederick Maynard**, (1905-1975), South African zoologist and Director of Pretoria's Transvaal Museum. His publications included *The lizards of southern Africa* (1943), *The snakes of southern Africa* (1962), and *A field guide to the snakes of southern Africa* (1970) which contained a number of line drawings by his brother, D.C. FitzSimons.

FLACCID PARALYSIS In reference to the effects of certain venomous snake bites, paralysis with the absence of muscle tone and symptomatic of secondary nerve damage.

FLAGELLUM (pl.) **FLAGELLA** A long whip-like outgrowth on micro-organisms, spores, gametes etc., that serves as an organ of locomotion and a term used, in herpetology, for the slender filament on the end of the tail of certain anuran larvae.

FLAME The fairly recently devised term applied to a striking crimson colour morph of the eastern garter snake *Thamnophis sirtalis sirtalis*.

FLASH COLORATION Referring to the very striking and contrasting colour on a normally CONCEALED SURFACE of many amphibians and reptiles, examples of which include the black and orange markings on the thighs and flanks of the otherwise green Ecuadorian leaf-folding frog (*Phyllomedusa tomopterna*) and the yellow, orange or red DEWLAP of *Anolis* lizards.

FLATUS DEFENSIVE STRATEGY A protective tactic demonstrated by certain snakes, e.g., the western hook snake *Gyalopion canum* and the Sonoran coral snake *Micruoides euryxanthus*, in which air is sucked into the CLOACA and then forced

78

out again under pressure. These defensive flatulent noises, which sound like deep squeaks, are sometimes accompanied with bubbles of faecal matter and act as a deterrent against predators.

FLATWORM Any free-living or parasitic invertebrate of the phylum PLATYHELMINTHES, which contains tapeworms and flukes. Their flattened bodies possess only one opening to the intestine and they are without a circulatory system. Many forms occur in both reptiles and amphibians.

FLEXUOUS Wavy; bending in a zigzag way.

FLICKING A particular activity seen in the aquatic *Xenopus*, in which both of the forelimbs are suddenly thrust forward together and then withdrawn a number of times in quick succession, and occurring whilst the amphibian is either suspended in the water or resting on the bottom. It serves to locate food, in the form of aquatic invertebrates which, when touched by the forelimbs, are quickly grasped and drawn into the mouth.

FLIPPER 1. Any one of the small, irregular-shaped pieces of skin cut from the legs of crocodilians in the commercial skin trade. 2. The flat, broad limb of sea turtles (family CHELONIIDAE), specialized for swimming.

FLOAT A collective noun for a group of crocodilians.

FLUKE Any parasitic flatworm of the class TREMATODA, which can occur in various internal organs of reptiles and amphibians.

FLUORESCENT LIGHTING A form of lighting often used to illuminate vivaria and emitted from a tube, coated with a thin layer of phosphor on its inner surface, in which an electrical gas discharge is maintained. In the husbandry of reptiles the gas is usually mercury vapour and emits ULTRAVIOLET RADIATION which causes the phosphor to fluoresce.

FLUVIAL Of, relating to, or occurring in, rivers and streams.

FOAM NEST A structure built by a number of different frog species, e.g., *Leptodactylus, Physalaemus,* in which to house their eggs and developing larvae. The nests of certain *Limnodynastes* species are built actually on the water's surface, whilst others, such as those of *Physalaemus* species, are constructed over small pools of water, suspended on a leaf. In American leptodactylids the JELLY is whipped into a frothy mass of foam by kicking movements of the hind legs of the female and/or male, whereas in Australasian myobatrachids it is created from bubbles produced by paddling movements of the female's front legs.

FOETAL WASTAGE In herpetology, used in reference to the loss of apparently viable eggs or FOETUS.

FOETUS The embryo in the later stages of development when its external features resemble those of the animal after birth. A term still used in herpetology but in its correct sense should be limited to embryos possessing an umbilical cord (i.e., mammals).

FOLD *See* DORSOLATERAL FOLD.

FOLIFORM Used to describe reptile scales that have an extensive free edge.

FOLLICLE A sac or small cavity occurring within a tissue or organ and having an excretory, secretory or protective function. Follicles within the ovary, for example, contain developing egg cells (ova).

FOLLICLE GLAND A pit-like structure located near the hindmost border of the VENTRAL scales of all crocodiles and gharials.

FOLLICULAR MATURATION The formation, in the GERMINAL tissues, and maturation of ovarian follicles, culminating in the release of ova which then enter the oviduct where they are fertilized if copulation occurs. The process of follicular maturation is closely linked to the FAT CYCLE.

FOLLICULAR VITELLOGENESIS The development of YOLK within an egg.

FOOD CHAIN The sequence of organisms, existing in a natural community, through which food energy is transferred, and expressed as feeding relationships in linear form. These interactions connect the various members of an ECOSYSTEM and illustrate how energy passes from one organism to another. Each organism, or link in the chain, obtains energy by eating the one below it and is eaten, in turn, by the one above it. For example, if a locust feeding upon a leaf is eaten by a frog, which is in turn eaten by a snake, which is then eaten by an eagle, the resulting food chain would be: plant-herbivore-insectivore-carnivore-carnivore. The eagle, in this example, is the top predator or tertiary consumer. Because many animals have more than one food source the idea of a 'food web' is perhaps more acceptable.

FOOD WEB *See* FOOD CHAIN.

FOOT STAGE The point in the development of anuran larvae, immediately following the PADDLE STAGE, in which the characteristic features of the foot become evident.

FOOT STAMP A particular form of behaviour occurring in certain lizards, in which the loser of a fight flattens its body to the ground and stamps its feet. This action indicates submission to its rival which then ceases its attack allowing the loser to take flight.

FORAGING Behaviour associated with the finding and devouring of food for which the animal must actively search or hunt.

FORAMEN Any small orifice, passage or perforation, such as that found in a bone through which blood vessels and nerves pass.

FORAMEN MAGNUM The large opening at the base of the skull which allows the connection between the spinal chord and the brain.

FORAMEN PANIZZAE An opening or partition (septum) near the base of the aorta of the heart in crocodilians, providing a passage between the left and right aortic arches enabling oxygen-rich blood to mix with blood poor in oxygen.

FORCE FEEDING In the husbandry of captive reptiles, the act of forcing a snake or lizard to swallow or eat food when the health and well-being of the particular animal is at risk, due to its continued reluctance to feed voluntarily for one reason or another.

FOREST-SAVANNAH MOSAIC A habitat type in which areas of forest are scattered with exposed regions of savannah.

FORM 1. Shape or size as distinct from texture, colour etc. 2. A vague term used in classification when the correct taxonomic rank is not clear. 3. Any distinct variety or MORPH occurring within a species. 4. A hollow or scrape in soil or amongst vegetation made by a terrestrial chelonian in which to sleep or shelter.

FOSSA An anatomical depression, shallow cavity or hollow area.

FOSSORIAL Burrowing; describing species that live largely below the soil or beneath ground vegetation, or their skeletons and limbs, which are adapted for digging and burrowing.

FOVEA The shallow depression on the retina of the eye, opposite the lens, present in reptiles (also in many mammals and birds) and characterized by numerous, slender and densely packed cones, giving greater image definition and better perception of colour.

FOVEAL Any one of the small scales bordering the LOREAL PIT of viperid snakes, which usually lie outside of the LACUNAL scales, over the SUPRALABIAL scales, beneath the LOREAL and directly in front of the eye. *See* POSTFOVEAL; PREFOVEAL.

FOVEATE Having small pits or depressions. **FRACTURE PLANE** *See* BREAKAGE PLANE.

FREEZE To become fixed, rigid or motionless; to remain in a fixed position for a length of time, especially through fear. Many reptiles and amphibians may freeze when suddenly disturbed or threatened.

FREEZE-DRYING; LYOPHILIZATION A process for preserving various organisms and substances, consisting of rapid freezing followed by desiccation in a vacuum. A technique used for preserving snake VENOM.

FRENOCULAR Commonly used for a scale situated between the LOREAL and the PREOCULAR scales or, in snakes, between the loreal and the eye.

FREQUENCY MODULATION In VOCALIZATION, the variation in pitch, over a length of time, of the CALL of a frog or toad.

FRIABLE Crumbly; easily broken apart or burrowed through.. Arid, sandy soils are often friable and are a habitat for a number of subterranean reptiles

FRICTION DISC *See* ADHESIVE PAD.

FRIGHT CRY The distress call of many anurans when grasped suddenly, in which a loud scream is emitted through an opened mouth, and readily distinguished from the typical anuran breeding call emitted with a closed mouth. The fright cry may well be successful in intimidating or startling some attackers but its aid as a defence against snakes, which are unable to hear air-borne sounds, is doubtful.

FRILL The enormous, free-edged fold of skin, covered with large keeled scales, located below and at either side of the head in the agamid frilled lizard *Chlamydosaurus kingii* of Australia and New Guinea. Normally folded back against the body, the frill is erectible, by means of a row of cartilaginous extensions ('ribs') of the hyoid apparatus, and extended during threat and courtship displays. The frill of an adult male may attain 300 millimetres in diameter.

FRINGE Any row of scales or spines extending from the body of reptiles but commonly used for the broad edging of scales on the digits of certain lizards (e.g., the genera *Acanthodactylus, Uma*) that enlarges the surface of the foot, aiding locomotion in loose sand.

FROGGERY A collective noun for a gathering of frogs.

FROG MALARIA The mosquito-transmitted disease known as 'amphibian malaria', caused by a protozoan, *Plasmodium bufonis*, which is a parasite of frogs and toads.

FROG TOXIN Any one of a number of poisonous substances, collectively known as batrachotoxins, secreted from the skins of many anurans (e.g., the families RANIDAE, HYLIDAE, DENDROBATIDAE and BUFONIDAE). In sufficiently high concentrations, frog toxins may prove fatal should they enter the bloodstream.

FRONTAL **1.** Pertaining to the scale, scales, or the space occupied by them (frontal area), on the top of the head situated between the SUPRAOCULAR scales, in chelonians, most lizards and snakes. **2.** A bone, narrow and prominent, on the outer surface of the skull roof, located between the eye sockets (ORBITS), usually separate from, but sometimes joined to, other bones such as the PARIETAL.

FRONTAL ORGAN *See* PARIETAL EYE.

FRONT-FANGED *See* ENDOGLYPH.

FRONTONASAL The scale, or scales, lying between the INTERNASAL, LOREAL and PREFRONTAL scales on the head in many lizards and chelonians, but rare in snakes. In snakes of the TYPHLOPIDAE family the frontonasal is one of a pair of scales lying above the NASAL and posterior to the ROSTRAL.

FRONTOPARIETAL **1.** The enlarged, either single or paired scale on the top of the head in chelonians, situated between the FRONTAL and the PARIETAL scales. **2.** In lizards, any one of the PLATES on the top of the head between the frontal and the parietal scales. **3.** In anurans, the bone resulting from the union of the frontal and parietal bones.

FRONTOPARIETAL FORAMEN
In certain anurans, the large opening
in the skull separating the paired
FRONTOPARIETAL bones.

FRONTOSQUAMOSAL ARCH In
certain urodeles, the bony arch which
unites the FRONTAL and
SQUAMOSAL bones.

FRUGIVORE Feeding on fruit.

FSC Acronym denoting the Field
Studies Council (United Kingdom).

FULVOUS Tawny-coloured;
reddish-yellow or reddish-brown.

FURCATE To divide into two equal
parts; fork. The tongues of all snakes
and some lizards are BIFURCATED.

FUSCOUS Dark-coloured; sombre
brownish-grey.

FUZZY A newborn mouse or rat that
has just started to grow its fur.

G

GAIT The manner of walking or running. For example, the leopard gecko *Eublepharis macularius* has a slow, stalking gait.

GALL BLADDER A small, pouch-like extension of the bile duct, occurring in most vertebrates, lying between the liver lobes and serving as a store for bile, which is produced in the liver and released in reaction to food entering the duodenum.

GALLERY FOREST Any forest along the banks of a watercourse in a region mostly free of other tree growth, and reliant upon the water table more than upon the climate as a whole.

GALLIWASP In the family ANGUIDAE, any lizard of the genus *Diploglossus,* inhabiting Central and northern South America.

GAMETE A mature, reproductive cell capable of uniting with another to form a new individual (ZYGOTE). Ova and spermatozoa are gametes.

GANGRENE The death and decay of tissue resulting from an interruption of the blood supply and a symptom in certain types of venomous snake bite.

GAPE The mouth-opening between the upper and lower jaws.

GAPING The condition, most frequently witnessed in snakes, in which the mouth is constantly opened, often for several minutes at a time, and often with the head and neck raised off the ground. The cause of gaping is almost always a respiratory infection in which the air passages become partially or totally blocked, making breathing difficult.

GARIAL An alternative spelling of GHARIAL.

GASTRALIA (sing,) **GASTRALIUM** Bony, rib-like rods located in the dermal layer of the skin of the abdomen, in crocodilians and the tuatara (*Sphenodon*). Gastralia, or 'abdominal ribs', are not connected to the vertebral column and lie outside the true ribs in the area between the pectoral and pelvic girdle.

GASTROENTERITIS An important and serious disease in reptiles, especially snakes, caused by parasitic protozoans of the genus *Entamoeba* and affecting the blood vessels and mucous membrane of the intestine. Unless treated early, it can spread throughout the infected reptile's body via the bloodstream with death resulting soon after.

GASTROLITH Any of the stones occurring in the stomachs of crocodilians, deliberately swallowed to grind up food in the absence of the ability to chew.

GASTROSTEGE Any one of the wide VENTRAL plates on the undersurface of most snakes.

GAVIAL An alternative name for GHARIAL.

GAVIALIDAE The gharial, *Gavialis gangeticus*. Family of the CROCODILIA, inhabiting deep

84

rivers of the Orient, especially the Ganges, Indus and Irrawaddy. MONOTYPIC.

GEKKONIDAE Geckos. Family of the SQUAMATA, suborder SAURIA, inhabiting the tropical, subtropical, and warmer temperate regions of the world. Thought to be now over 1000 species in some 80 or more genera. Geckos can be largely divided into two key groups: those species which have fixed eyelids i.e., the Gekkoninae, Diplodactylinae, and Teratoscincinae, and those which have closable eyelids i.e., the Eublepharinae and Diplodactylinae. Those in the former group can be further separated into primarily arboreal species with sticky footpads, and mainly terrestrial species which lack footpads.

GENE A unit of heredity, composed of DNA, carried in the chromosome transmitted from generation to generation by the GAMETES and governing the development and characteristic features of an individual. In the captive breeding of reptiles and amphibians, it is necessary to periodically replenish the GENE POOL to avoid INBREEDING depression

GENE FIXATION The establishment of a GENE once it becomes present in every individual of a given population.

GENE POOL The total number and variety of the genes of a given breeding population or species existing at a given time.

GENERALIST An animal having a diet consisting of a variety of foods.

GENERATION TIME The amount of time necessary for a generation of hatchlings or newborn (NEONATES) to become sexually mature and produce their own young. The generation time for young Indonesian blue-tongued skinks *Tiliqua gigas*, for example, is 12 months under correct conditions.

GENERIC Of, relating to, or a member of a GENUS.

GENETIC Of, or relating to, heredity or the features and traits acquired by the offspring from their parents. The blueprints of inherited characteristics are passed on to the young in GENES, a particular genetic arrangement being termed a GENOTYPE.

GENETIC DIVERSITY *See* BIODIVERSITY.

GENIAL A term frequently used by early authors for the CHINSHIELD in snakes and lizards.

GENIOLABIAL A scale, in alligator lizards (genus *Gerrhonotus*), situated between the INFRALABIAL and CHINSHIELD scales.

GENOTYPE The genetic composition of an organism, or a group of organisms all having the same genetic composition.

GENTAMYCIN An antibiotic useful in reptile husbandry for the treatment of a wide range of bacterial diseases and particularly successful against GRAM-NEGATIVE BACTERIA.

GENUS (pl.) **GENERA** A taxonomic category used in the classification of organisms, consisting of a group of species having similar structural characteristics and ranking below

FAMILY and above SPECIES. The common name of an organism is sometimes identical or similar to that of the genus, e.g., viper, *Vipera*; chameleon, *Chamaeleo*.

GERMINAL Of, or relating to, a germ or a germ cell, or relating to the embryonic stage of development.

GERMINAL EPITHELIUM Embryonic tissue consisting of one or more sheets of closely packed cells that cover the internal and external surface of the body in the earliest stages of growth.

GESTALT A perceptional pattern or formation having merits as a whole that cannot be expressed simply as a sum of its parts; a German word employed in natural history to describe the broad entire appearance of an animal; an animal's JIZZ.

GESTATION The period of development, or carrying of embryos, by the female of an animal exhibiting VIVIPARITY, from the fertilization of the egg to the birth (PARTURITION) of the young.

GHARA The prominent, fleshy hump that appears on the tip of the snout of mature gharials, and which develops to form a lid covering the nostrils. During the male's territorial patrols, or when courting, the lid flaps as the reptile exhales producing a loud buzzing noise that serves as a warning to competitors. This cartilaginous nose-knob is named 'ghara' from the Hindu term for 'mud pot', and it is this feature that gives the species its common name, GHARIAL.

GHARIAL; GARIAL; GAVIAL A large, monotypic, fish-eating Indian crocodilian *Gavialis gangeticus* with a very long slender snout. Should not be confused with the false gharial *Tomistoma schlegelii*

GHOST The term given to a colour morph which is both HYPOMELANISTIC and ANERYTHRISTIC.

GIBBOSE Having a hunchback; humpbacked.

GIBBOSITY A bulge or protuberance.

GILA MONSTER *See* HELODERMATIDAE.

GILL A respiratory organ used by aquatic animals to effect the exchange of respiratory gases between the animal's blood or body fluids and the water in which it lives. Located in the region of the neck, gills may be feathery, as in urodele and young anuran larvae, or enclosed, as in older anuran larvae. *See* ALLANTOIC GILL; EXTERNAL GILL; INTERNAL GILL; OPERCULAR GILL.

GILL ARCH Any one of the curved cartilaginous or bony bars situated one behind the other and extending dorsoventrally on each side of the pharynx, supporting the amphibian gills

GILL RAKER Any one of the cartilaginous or bony filaments on the inside of a GILL ARCH in amphibian larvae, which serve as filters and prevent solid substances from being carried through the GILL SLITS.

GILL SLIT In certain amphibians and amphibian larvae, any one of the series of paired, narrow openings at the base of the external gills through which water can flow from the sides of the pharynx to the outside.

GINGIVAL SULCUS A small indentation at the base of a tooth where it enters the gum.

GIRARD, Charles Frederic, (1822-1895), a Swiss zoologist who worked at the Smithsonian Institution, Washington DC, with Spencer Fullerton Baird, with whom he co-authored many publications, including the *Catalogue of North American reptiles* (1853).

GIRDLE 1. The ring of bones situated anteriorly (PECTORAL GIRDLE) and posteriorly (PELVIC GIRDLE) in the trunk of the TETRAPOD body, supporting the arms (forelimbs) and legs (hind limbs) respectively. **2.** In the commercial crocodilian skin trade, the term given to any of the pieces skinned from the belly and thighs directly in front of the vent and most frequently taken from large caimans, e.g., *Melanosuchus.*

GIRTH The measurement around a reptile or amphibian, usually at the abdomen or, in snakes, at mid-body. Frequently incorrectly given as the diameter but the true girth of an animal is its circumference.

GIZZARD The thickened, muscular stomach of crocodilians, often found to contain GASTROLITHS which aid in digestion.

GLAND Any one of a number of cells, occurring either within special organs or individually in the epithelium, that synthesize chemical substances and secrete or excrete them directly into the bloodstream or through a duct, e.g., oil, wax, saliva, venom etc.

GLANDULAR Of, relating to, containing, functioning as, or affecting a GLAND.

GLANDULAR SCALE The modified, secretory scale in certain lizards, situated anterior to the CLOACA or on the thigh.

GLIDE The method of unpropelled movement through the air used by certain arboreal reptiles (e.g., flying lizards of the genus *Draco*) and amphibians (e.g., flying frogs of the genus *Rhacophorus*) in their progress from one tree to another or to the ground, in which the body or appendages are flattened or spread and act as a gliding surface.

GLOBAL ZOO DIRECTORY A comprehensive register of zoological collections worldwide detailing addresses and telephone numbers. Many entries also include numbers of animals held, staff size and attendance figures.

GLOBULINS Blood plasma proteins with a number of roles such as hastening the exchange of prothrombin to THROMBIN, and antibody action.

GLOSSAL SKELETON *See* HYOID APPARATUS.

GLOSSOHYAL A muscle of the extensible tongue.

GLOSSOPHARYNGEAL NERVE The ninth cranial nerve which supplies the muscles of the

PHARYNX, the tongue, the middle ear, and the PAROTID GLAND. It permits swallowing and palatal movement

GLOTTIS The opening from the PHARYNX to the TRACHEA (windpipe), the front section of which, in snakes, is protruded during the act of swallowing large prey items and therefore prevents the reptile from suffocating.

GLOVE; NUPTIAL GLOVE An area of densely packed, hook-like processes on the under surface of each forelimb in male *Xenopus* frogs, used to maintain a firm hold on the females during AMPLEXUS.

GLUED AMPLEXUS A form of AMPLEXUS in which the males in certain anuran species attach themselves to the backs of the females by means of a sticky secretion produced in abdominal glands. Occurs in a number of plump, short-limbed species, including members of the genera *Breviceps, Gastrophryne* and *Kaloula.*

GLYCOGEN In animals, the stored form of carbohydrate.

GLYPHODONT Having teeth, which are grooved, through which VENOM flows into the tissues of a bitten animal; having FANGS; glyphous.

GLYPHOUS *See* GLYPHODONT.

GOANNA An Australian term for a monitor lizard (*Varanus* sp.), and derived from the name 'iguana'.

GOITRE A disturbance in the development, growth and function of the thyroid gland caused, usually, by an iodine-related reduction in hormone secretion. Amphibian larvae with thyroid disease frequently double or even treble in size, fail to metamorphose and die. In reptiles, a goitre often results in an enlargement of the thyroid gland, appearing as a hard, fibrous growth in the region of the throat.

GONAD One of the usually paired reproductive organs responsible for the production of GAMETES, such as the male TESTIS (producing spermatozoa) and the female OVARY (producing egg cells, or ova).

GONADAL RECRUDESCENCE The formation of spermatozoa in male testes after a period of dormancy. In many reptiles, especially temperate species, it occurs after or during a period of warming which has been preceded by a period of cooling (HIBERNATION).

GONADAL SUPPRESSION The failure of spermatozoa to develop in the GONADS which, in reptiles, is usually due to prolonged exposure to temperatures below the lower limit of their optimal thermal range, resulting in infertile mating.

GOUT A condition, sometimes affecting reptiles, in which urates normally expelled from the body via the kidneys accumulate in the tissues, joints (uratic arthritis) or internal organs (visceral gout). The deposition of urate crystals destroys the internal organs so that they can no longer function correctly and, untreated, can result in death.

GRAB-STICK Any instrument used to grasp and restrain snakes and lizards, particularly aggressive or venomous species. Such sticks usually consist of a strong pole or rod with a grasp arrangement at one end and a trigger for opening and closing the grasp at the other.

GRAM-NEGATIVE BACTERIA One of two major groups of bacteria commonly found in reptiles and amphibians and including those of the *Aeromonas* and *Pseudomonas* genera. *See* AEROMONAS INFECTION; PSEUDOMONAD INFECTION; SEPTICAEMIA.

GRANULAR Possessing a surface with a texture resembling granules or minute grains.

GRANULAR GLAND One of several glands occurring in the skin of amphibians, producing a toxic secretion that can cause much irritation and damage to mucus membranes and which, in some species of *Bufo,* can be squirted some distance. Granular glands are frequently enlarged and arranged close together, e.g., the PAROTOID in members of the BUFONIDAE family of toads.

GRANULOMA A tumour-like mass resulting from a chronic inflammatory or infectious condition.

GRASSLAND A region in which the climate is hot and dry for long periods during summer and below freezing during the winter. Grasslands, which first appeared in the MIOCENE, are usually characterized by grasses and other upright herbs, and a lack of trees or shrubs, and occur in the dry temperate interiors of continents.

GRAVID The term used, in herpetology, to describe the condition of a female carrying eggs or young, from fertilization to the moment of birth or, in species exhibiting OVOVIPARITY, from fertilization to the birth of the young still enclosed in their egg membranes.

GRAY, **John Edward,** (1800-1875), British zoologist at the British Museum (Natural History), London, he produced a number of catalogues, among them the *Catalogue of shield reptiles* (1855 & 1870), relating to the collections held at the BM (NH), which included many descriptions of new taxa.

GRESSORIAL Applied to limbs that are modified for running.

GROIN The slight depression or hollow at the points where the hind limbs join the abdomen.

GROOVE A long, narrow channel or furrow on a bodily structure or part, e.g., the COSTAL FOLD in urodeles, and the groove in the teeth of many reptiles, liked with the conduction of venom.

GROUND COLOUR The usually plain-coloured background on which other colours, patterns or markings are overlaid.

GROWTH RING; ANNUAL RING Any one of the number of concentric, ring-like areas (annuli) occurring on the laminae of many chelonians, most obvious in juveniles, and each one said to represent one year's growth.

GUANOPHORE A pigment cell of the skin containing white or blue crystals of guanine, a substance related to uric acid.

GULAR On, or pertaining to, the region of the throat. Also used for the foremost laminae (paired or single) on the PLASTRON of most chelonians.

GULAR CREST In certain lizards, a row of elevated scales beneath the chin.

GULAR DISC A distinct, circular area of skin in the centre of the throat of many male anurans. Thicker than the skin surrounding it, the gular disc, which overlies the VOCAL SAC, is expanded as the throat inflates when the amphibians CALL.

GULAR EXPANSION *See* DEWLAP

GULAR FAN *See* DEWLAP.

GULAR FOLD A transverse fold of skin across the rear of the throat directly in front of the point where the forelegs join the abdomen. The fold is well-developed in many urodeles and in some lizards, where its presence or absence, and the scales upon it, is taxonomically significant.

GULAR SAC An area of expandable skin on the throat of many male anurans, which inflates like a balloon when the animal calls; the VOCAL SAC of frogs and toads.

GULAR TENTACLE An alternative term for BARBEL, one of the fleshy projections occurring on the chins and throats of some chelonians.

GULAR VIBRATION In salamanders, the rapid movement of the BUCCAL floor, possibly aiding respiration but probably assisting olfaction (the sense of smell).

GÜNTHER, Albert Carl , (1830-1914), German herpetologist and ichthyologist at the British Museum (Natural History), London. As well as contributing several publications on fish, his many herpetological contributions included catalogues of the collections held at the BM (NH), including the tailless amphibians (1858), colubrine snakes (1858) and the giant land tortoises (1877). He wrote the classic *Reptiles of British India* (1864), which provided the basis for all modern research on the reptiles and amphibians section of the *Biologica Centrali-Americana* (1885-1902), which is considered one of the most significant works on the herpetology of Central America.

GUT-LOADING The practice of feeding prey items, for example crickets and mealworms etc, a vitamin and mineral supplement or good quality fruit and vegetable mix in order to fulfil the nutritional requirements of captive reptiles and amphibians feeding on these prey items.

GYMNOPHIONA The limbless amphibians - CAECILIIDAE, or APODA, an order of tropical worm-like animals of the class Amphibia.

H

HABITAT The particular region, characterized by certain features such as vegetation, climate and topography etc., where an animal or plant lives.

HABITAT DIVERSITY *See* BIODIVERSITY.

HABITUS General physical condition or appearance.

HABU Any one of the Asiatic PIT VIPERS of the genus *Protobothrops*, the largest of which is the Okinawa habu. *Protobothrops flavouridis* of the Ryukyu Islands, which reaches some 200 cm in length. Others include the Chinese habu *Protobothrops mucrosquamatus*, and the Sakishima habu *Protobothrops elegans*.

HAEMATEMESIS The vomiting of blood and a symptom in certain cases of venomous snake bite.

HAEMATOCRYAL An alternative description for a POIKILOTHERM.

HAEMATOLOGICAL Relating to the blood and blood-forming tissues.

HAEMATOTOXIC *See* HAEMOTOXIC.

HAEMATURIA In reference to the effects of certain venomous SNAKE BITES, the passage of blood in the urine.

HAEMOCOAGULIN A substance, common in snake venoms, hastening or slowing down the coagulation of the blood.

HAEMOGLOBIN A complex protein that gives red blood cells their characteristic colour and which carries oxygen to the tissues.

HAEMOGLOBINAEMIA; HAEMOGLOBINURIA Passage of the blood pigment HAEMOGLOBIN in the urine giving a pinkish, reddish or blackish colour depending on the quantity of haemoglobin present.

HAEMOLYSIN In reference to SNAKE BITE, a constituent of a venom (a HAEMOTOXIN) which destroys red blood corpuscles, resulting in haemoglobin (the oxygen carrying element of the blood) being lost in the urine which consequently turns red (haemoglobinuria).

HAEMOLYSIS The obliteration of red blood cells resulting in the liberation of HAEMOGLOBIN into neighbouring fluids.

HAEMOLYSIS RATE The rapidity with which a HAEMOLYSIN will cause red blood cells to collapse.

HAEMOLYTIC *See* HAEMOTOXIC.

HAEMORRHAGIC Said of any constituent of certain snake venoms which causes internal bleeding into the tissues and ensuing blood loss resulting from blood vessel damage.

HAEMORRHAGIN Any toxic protein that destroys the walls of thin blood vessels resulting in bleeding.

HAEMOSTATIC DISTURBANCE; HAEMOSTATIC DISRUPTION In reference to venomous SNAKE BITE, defective clotting of the blood resulting in excessive bleeding.

HAEMOTOXIC; HAEMATOTOXIC, HEAMOLYTIC, HEMOTOXIC Describing the effects of certain snake-venom components which destroy blood vessels and prevent the blood from functioning. *Compare* NEUROTOXIC.

HAEMOTOXIN Any constituent of snake venom that results in a breakdown of the blood cells, causing internal haemorrhaging and localized bruising.

HALF-SIB MATING A mating between individuals which have the same parents.

HALLUX The first (innermost) digit on the hind foot of most four-limbed reptiles, and amphibians (and other tetrapods).

HALTERE *See* STABILIZER.

HAMADRYAD The king cobra *Ophiophagus hannah*, the longest venomous snake in the world, inhabiting India and South-East Asia.

HAMATE Hook-shaped.

HARDERIAN GLAND In reptiles, one of the two large glands connected with each eye which produce the 'tears' which help to keep the eye moist. Associated with the NICTITATING MEMBRANE in crocodilians, chelonians and lizards. *See* LACRIMAL GLAND.

HARDUN *Laudakia (Agama) stellio,* a lizard of the family AGAMIDAE, occurring in rocky habitats in north-eastern Africa to south-western Asia and parts of Greece.

HASTATE Having a pointed tip and two outward-pointing lobes at the base; arrow-shaped with the rear tips of the arrowhead extended and curved outward.

HCT Acronym denoting the HERPETOLOGICAL CONSERVATION TRUST.

HEAD AND TRUNK LENGTH In reference to the embryos of alligators, the space along the middorsal line between the tip of the snout and the base of the hind limb.

HEAD-BODY LENGTH The length of a caecilian, anuran, crocodilian, snake or lizard, measured along the centre of the body from the tip of the snout to the vent; the SNOUT-VENT LENGTH.

HEAD CAP The area of black coloration that covers the top of the snout in coral snakes (*Micrurus* and *Micruroides*), extending backwards frequently to spread over most or all of the PARIETAL scales on the top of the head.

HEADSLAP The action performed by mature crocodilians in which the head is first lifted above the water so that the lower jaw is just visible. This position is frequently held for several minutes before the reptile rapidly opens and closes its jaws as if biting at the water, creating a sudden audible 'pop', followed by a tremendous splash. The display, consisting of the headslap, JAWCLAP and related behaviour, serves to attract females during courtship and acts also as a deterrent to rival males.

HEATH A lowland area typical of parts of Europe, with a sandy, acid and infertile soil, dominated by heather *Calluna vulgaris*, with scattered trees and bushes, notably gorse *Ulex europaeus.* Good examples are the sandy heaths of southern and western England, many of which contain interesting and, in some cases rare, reptiles and amphibians.

HEAT ROCK *See* HOT ROCK.

HEAT SENSORS *See* LABIAL PIT; LOREAL PIT.

HEDONIC Having pleasurable sensations and, as used in herpetology, acting as a stimulus to sexual activity.

HEDONIC GLAND Any gland, situated on various parts of the head and tail in certain urodeles, the secretion of which induces sexual excitement or stimulation in the females when the male rubs it against the female's body.

HEIFER A subadult female terrapin and, less commonly, crocodilian.

HELEOPHRYNIDAE Ghost frogs. Family of the ANURA, inhabiting South Africa. Three species in one genus.

HELIOTHERM Any animal that needs to bask in the sun in order for THERMOREGULATION to take place. Reptiles are heliothermic.

HELMET The bony structure on the top and back of the skull of certain lizards, especially *Basiliscus.*

HELMINTH Any one of a miscellaneous group of parasitic worms and flukes inhabiting the respiratory, circulatory or digestive system of reptiles and amphibians. Examples are the phylum ACANTHOCEPHALA and classes NEMATODA, TREMATODA and CESTODA. Death from helminthiasis may occur in animals with heavy burdens of these parasites.

HELMITHIASIS Infestation with parasitic worms, or HELMINTHS.

HELODERMATIDAE Beaded lizards. Family of the SQUAMATA, suborder SAURIA, inhabiting south-western North America and parts of Central America, and the only venomous lizards in the world. The family has one genus, *Heloderma,* containing two species *Heloderma suspectum,* the gila monster, and *Heloderma horridum,* the Mexican beaded lizard. Beaded lizards have no hereditary link to snakes from which they differ by having their 'fangs' and venom glands located in their lower jaw.

HEMIPENIS (pl.) **HEMIPENES** The grooved copulatory organ of male snakes and lizards, paired and located within the base of the tail behind the cloaca. The surface may be covered with an array of hooks and spines which help to hold the organ securely in place in the female's cloaca during copulation.

HEMOTOXIC *See* HAEMOTOXIC.

HENOPHIDIA A group of snakes which diverged from the ALETHINOPHIDIA and includes, amongst many families, the pythons, boas, and dwarf boas. Henophidians possess vestiges of a pelvic girdle and teeth on the premaxillae. Taxonomy of snakes within the group remains debatable and is being constantly reviewed.

HENOPHIDIAN *See* HENOPHIDIA.

HEPARIN A substance that hinders coagulation by averting conversion of prothrombin to THROMBIN.

HEPATIC Of, or relating to, the liver.

HEPATOTOXICITY The breakdown of the liver due to the effects of a TOXIN.

HERBIVORE An animal that feeds upon grass and plant matter. Many terrestrial chelonians are herbivorous.

HERMAPHRODITE An animal that possesses both male and female organs of reproduction.

HERP An abbreviation for HERPETOLOGY, but used frequently for any individual amphibian or reptile.

HERPES A highly contagious viral disease, commonly known as 'grey spot' occurring mainly in freshwater chelonians, especially juveniles. Herpes viruses are also found in marine turtles, iguanid lizards and many snakes.

HERPET. An abbreviation for HERPETOLOGY.

HERPETOCULTURE The breeding of reptiles and amphibians in captivity.

HERPETOFAUNA The term used to embrace all the reptile and amphibian populations of a particular area, region or country.

HERPETOL. An abbreviation for HERPETOLOGY.

HERPETOLOGICAL CONSERVATION TRUST; HCT A United Kingdom based charitable organization The Herpetological Conservation Trust was established in 1989 to safeguard Britain's threatened amphibians an reptiles by: protecting, managing, and improving the actual sites inhabited by these rare and threatened animals; increasing the knowledge of their way of life and habitat requirements through research and education, and by providing expert advice; and raising public awareness in order to stimulate interest in and understanding of our amphibians and reptiles.

HERPETOLOGIST A person who studies reptiles and amphibians.

HERPETOLOGY The study of reptiles and amphibians, including their anatomy and classification (TAXONOMY), history and behaviour, and embracing ecology, biochemistry, immunology, zoogeography, toxicology and computer science.

HERPETOZOA A group expression for both amphibians and reptiles.

HERPTILE Any individual amphibian or reptile. Considered by some to be slang and therefore inappropriate in academic literature, the term is used widely nevertheless by both amateur and professional authors.

HERPTILIARY; REPTILIARY An outside, open-air enclosure used to house amphibians and/or reptiles in near natural conditions.

HERTZ A measure of sound frequency in cycles per second; low numbers represent the low or bass sounds; high numbers, high treble or

94

pitched sounds.

HETERO- A prefix meaning 'different from'.

HETERODONT Having teeth of several types, such as molars and canines, for different purposes such as crushing, grinding, tearing and cutting etc. *Contrast* HOMODONT.

HETEROGENEOUS Lacking pure genes for a characteristic feature. For example, a reptile or amphibian that is heterogeneous for albinism has the characteristic coloration typical of its species but carries one gene for albinism.

HETEROGENOUS Having a colour, pattern or size that is not uniform. *Compare* HOMOGENOUS.

HETEROLOGOUS ANTIVENIN In reference to ANTIVENIN prepared from sources other than the species causing the envenomation.

HETEROMORPHOUS Differing in some way, e.g., size, shape, function etc., from that which is usually considered normal for a particular species.

HIBERNACULUM (pl.) **HIBERNACULA** The winter retreat of a hibernating animal. The location in which an amphibian or reptile hibernates, or remains inert during the colder part of the year. The hibernaculum may be a disused rodent burrow, a mine shaft, in mud, beneath the roots of trees, or under decomposing debris etc., where humidity and temperature remain relatively constant.

HIBERNAL Pertaining to the winter. *Compare* AESTIVAL; AUTUMNAL; PREVERNAL; SEROTINAL; VERNAL.

HIBERNATION An extended period of rest, torpor or inertness during the winter months, in which metabolism is greatly slowed down. Hibernation occurs in most temperate reptiles and amphibians and is frequently very prolonged in the most northerly species.

HIDE BOX In captive reptiles and amphibians, any object which the animal may use as a place of REFUGE, for example: a hollow log; a wooden box; or an upturned flowerpot etc.

HIERARCHY The dominance-submission interrelationship (peck order), between each successive grade or rank established within a population of animals, with the most dominant member holding its position over the lower, more submissive grades by periodic combat or threat displays and allowed priority in access to females, food etc., by subordinates. Such a hierarchy has to be carefully planned and observed in a mixed-sex community of lizards if serious injuries, and even fatalities, are to be avoided amongst captive animals.

HIGH KICKING In frogs of the *Xenopus* genus, the action presented by an individual suddenly, and often repeatedly, drawing its hind limbs forward until its toes reach the level of its shoulders whilst pulling its body backward followed by it swiftly retracting its limbs. Behaviour frequently accompanied by FLICKING.

HIME HABU *Ovphis okinavensis*, a highly venomous Asiatic PIT VIPER, of the Ryu Kyu Islands of Japana. *See* HABU.

HINGE The broad, flexible joint in the shell of certain chelonians that allows the front or rear of the shell to close, enabling the reptile to seal itself within when threatened. Examples of the hinge occur in the mud turtle (*Kinosternon*) in the PLASTRON, and in the hingeback tortoise (*Kinixys*) in the CARAPACE, amongst others.

HISPID Covered with stiff hairs or bristles; possessing a coarsened, bristly surface as a result of minute spines.

HISTAMINE An amine derived from histidine and released by the body tissues in many allergic reactions. It also stimulates the secretion of gastric juices, dilates blood vessels and contracts muscles. *See* ANTIHISTAMINE.

HOLARCTIC The largest zoogeographical region comprising the PALAEARCTIC (Old World) and NEARCTIC (New World) regions, and covering most of the northern hemisphere. Faunal groups from these two regions are basically similar.

HOLBROOK, **John Edward,** (1794-1871), American medical practitioner and herpetologist whose *North American herpetology* became the basis of American herpetology when it was published as a five-volume, second edition in 1842. The first edition was produced in three volumes between 1836 and 1838.

HOLO- A prefix meaning whole; entire.

HOLOCENE Pertaining to the second and most recent epoch of the Quaternary period. *See* RECENT.

HOMEO-; HOMOEO-; HOMOIO- A prefix meaning resembling or similar to.

HOMEOSTASIS The regulation of a metabolic balance within an animal, such as the maintenance of body temperature. It enables cells to function more efficiently and any factor that disrupts, or threatens to disrupt, homeostasis is known as a 'stressor'. STRESS in captive reptiles or amphibians is any disruption of the normal physiological processes of the animals' internal environment, or homeostasis.

HOMEOTHERM An alternative spelling for HOMOIOTHERM.

HOME RANGE The area or areas which an individual animal occupies during its life, including both permanent and seasonal territories, together with the tracks and corridors through which it moves. Not to be confused with the DISTRIBUTION of the SPECIES as a whole.

HOMING Pertaining to an animal's ability to return home, frequently after having travelled great distances. Particularly prevalent in many amphibians, the homing instinct enables them to return to breed in the same waters where they themselves were spawned.

HOMO- A prefix meaning 'same'.

HOMOGENOUS Having a colour, pattern or size that is uniform.

Compare HETEROGENOUS.

HOMODONT Having teeth which are all of a single type with no difference from one part of the jaw to the other, and occurring, for example, in anurans in which all the teeth are correspondingly small, conical prominences on the MAXILLA, PREMAXILLA, and VOMER bones of the upper jaw. *Contrast* HETERODONT.

HOMOGENEOUS;
HOMOGENOUS Of the same kind; uniform, and used in herpetology to describe the state in which scales on a reptile correspond in shape and size.

HOMOIOTHERM;
HOMEOTHERM Warm-blooded; an animal e.g., a bird or mammal, that maintains its internal body temperature at a constant level by using metabolic processes to override fluctuations in the temperature of the environment. The extinct pterosaurian reptiles are thought to have been homoiothermic (or homoiothermal) but all RECENT reptiles and amphibians are cold-blooded, or POIKILOTHERM.

HOMOIOTHERMIC;
HOMOIOTHERMAL *See* HOMOIOTHERM.

HOMOLOGOUS Used in comparative anatomy to refer to features having the same evolutionary origin but not necessarily the same function. The forelimbs and hind limbs of all terrestrial vertebrates are built on the same five-digit (PENTADACTYL) arrangement and are said, therefore, to be homologous.

HOOD The flattened skin expansion of the back of the neck region in certain erect cobras.

HOODIE Herpetologists' slang for any of the cobras that spread a HOOD when alarmed.

HOPLOCERCIDAE Whorl-tailed and wood lizards etc. FAMILY of the SQUAMATA, suborder SAURIA, formerly of the IGUANIDAE, inhabiting Central and South America from Panama to Peru. Some 10 species in three genera.

HOPPER 1. The name given to individual sub adult mice, on account of their habit of leaping in the air when pursued. **2.** A hatchling locust, cricket or grasshopper.

HORIZONTAL UNDULATORY MOVEMENT *See* SERPENTINE MOVEMENT.

HORN; CORNU Used in herpetology for any, usually pointed, epidermal projection on the head of an amphibian or reptile, such as the enlarged, elongated scales on the snout of the African vipers *Bitis gabonica* and *Bitis. nasicornis*, and above the eyes in *Cerastes cerastes*. In amphibians, examples include the fleshy excrescences above the eyes in the frogs *Ceratophrys* and *Megophrys nasuta*.

HORNBACK In the commercial skin trade the rough, dorsal skin of a crocodilian, in the centre of which are the large, bony dorsal scales, normally possessing raised keels, and bordered on both sides by the smoother, almost square shaped scales of the ventral surface.

HORNED TOAD The somewhat misleading name given to the swollen- and flat- bodied iguanid lizards of the genus *Phrynosoma*, such as *P. cornutum* of Texas for example.

HOST The organism in, or on, which another organism (the PARASITE) lives and derives nourishment. If a snake is parasitized by a tapeworm, for example, the snake is the host.

HOT-ROCK; HEAT ROCK A commercially-available, purpose-built artificial stone or rock, containing a heating element which emits heat, for use in a vivarium as a supplementary source of warmth. Such products are, unfortunately, notoriously unreliable and frequently the cause of THERMAL BURNS in reptiles which come into contact with them.

HUMERAL The second foremost pair of laminae, situated between the GULAR and the PECTORAL laminae on the plastron of chelonians.

HUMERUS The long bone of the upper forelimb in all tetrapod vertebrates. Its upper, rounded head articulates with the SCAPULA, and the lower end with the RADIUS and ULNA, at the elbow.

HUMIDITY Relating to dampness, or the measure of the moisture content in the air. Humidity is an important factor in the successful keeping and breeding of reptiles and amphibians, with some species requiring almost none at all and others depending on it completely for their survival.

HUMP The natural, rounded protuberance on the lower back in anurans, where the vertebral column joins with the pelvic girdle.

HUSBANDRY Every aspect of the care and management of both domestic and captive wild animals.

HYALURONIDASE In reference to the effects of certain venomous snake bites, an enzymatic protein that advances the local spread of a toxin by destroying hyaluronic acid, the cement substance of tissues.

HYBRID The often sterile offspring resulting from the pairing of a male of a certain species, genus or race etc., with the female of another.

HYBRIDIZATION Interbreeding between individuals of different species that differ genetically in at least one characteristic, the product of such a cross being called a HYBRID or INTERGRADE. Only species that are related can hybridize, the offspring from such unions being either similar in appearance to one of the parents or intermediate to both, depending on whether a particular inherited characteristic is dominant or recessive.

HYBRID ZONE A geographical zone in which the HYBRIDS of two geographical races may occur.

HYDRIC Used in reference to areas or regions that are wet or swampy.

HYDROPHIIDAE Sea snakes. Family of the SQUAMATA, central order Caenophidia, inhabiting the Indo-Pacific and neighbouring tropical and subtropical seas. 46 species in 16 genera.

HYDROPS In anurans, a condition in which excess fluid collects in one or more of the lymph sacs lying beneath the skin. Captive anurans seem particularly susceptible to hydrops, as do the larvae of females which have had oviposition induced artificially.

HYDROTAXIS The movement of an organism in response to the stimulus of water, an example of which occurs in hatchling marine turtles which make straight for the sea on emerging from the sand, even though they may not be able to see it.

HYGROMETER An instrument for measuring humidity.

HYGROMETRY The measure of humidity with an instrument or gauge.

HYGROSCOPIC Able to absorb moisture readily from the air, as do many reptile eggs which absorb moisture throughout their incubation and must therefore be deposited in a humid environment.

HYLIDAE Tree frogs. Family of the ANURA, suborder Procoela, inhabiting most tropical and subtropical parts of the world except Africa. Approximately 600 species in over 27 genera.

HYOBRANCHIAL APPARATUS Group of muscles and bones which, in urodeles, enables the tongue to protrude from the mouth; hyoglossum; hyoid; hyoid apparatus.

HYOGLOSSUM *See* HYOBRANCHIAL APPARATUS.

HYOID *See* HYOBRANCHIAL APPARATUS.

HYOID BONE A horseshoe-shaped bone which supports the tongue in tetrapod vertebrates. In some lizard species, such as *Anolis*, the bone is movable and used to extend and strengthen the throat pouch during courtship and territorial displays.

HYOPLASTRON (pl.) **HYOPLASTRA** Either one of the second foremost pair of bones in the PLASTRON skeleton of a chelonian.

HYPAPOPHYSIS (pl.) **HYPAPOPHYSES** A usually single, spine-like process on the vertebrae in some snakes and lizards. In egg-eating snakes (*Dasypeltis*), the tips of the hypapophyses in the oesophagus are used to break the shells of eggs as they are swallowed.

HYPERCOAGULABILITY In reference to the effects and treatment of certain VENOMOUS snake bites, the coagulation of blood at an accelerated rate resulting in excessive clotting.

HYPEROLIIDAE Family of the ANURA, inhabiting Africa and Madagascar, with the exception of one species, *Tachycnemis seychellensis*, which occurs on the Seychelle Islands. 193 species in 14 genera.

HYPERSENSITIVITY Having an abnormal sensitivity to an allergen or other agent such as a snake VENOM.

HYPERTHERMIA A condition in which the body temperature is dangerously high. Reptiles and amphibians will die if exposed to temperatures above their VOLUNTARY MAXIMUM, which varies from one species to another but lies only a few degrees above the upper limit of their normal activity range.

HYPERVITAMINOSIS A nutritional disorder found mostly in captive reptiles and caused by an excess of vitamins in their diet. Too much vitamin D3, for example, can cause metabolic disturbances of the bones.

HYPNOTIC REFLEX An alternative term for DEATH FEINT.

HYPOCALCEMIA A deficiency of body calcium.

HYPOMELANISM A condition in which black pigmentation is significantly reduced but not absent. True hypomelanistics are a result of a recessive genetic mutation.

HYPOMELANISTIC *See* HYPOMELANISM.

HYPOPLASTRON (pl.) **HYPOPLASTRA** Either one of the second hindmost pair of bones in the PLASTRON skeleton of a chelonian.

HYPOTENSION Low blood pressure.

HYPOTHENAR TUBERCLE The small knob-like projection situated beneath the base of the outermost (fourth) digit of an anuran forefoot.

HYPOTHERMIA A condition in which the body temperature is dangerously low. Reptiles and amphibians will become torpid as the ambient temperature is lowered and if it continues to fall below their VOLUNTARY MINIMUM death will follow.

HYPOVITAMINOSIS; AVITAMINOSIS A nutritional disorder that can develop in captive reptiles and amphibians, resulting from a failure to provide sufficient or correct vitamins in their diets. Swollen eyes and lack of appetite in chelonians due to a lack of vitamin A, and convulsions in certain fish-eating snakes due to a lack of vitamin B1, are both examples of hypovitaminosis.

I

ICHTHYOPHIIDAE Caecilians. Family of the GYMNOPHIONA, inhabiting tropical South America, South-East Asia and the Indo-Australian islands. 43 species in 4 genera.

ICONOTYPE An illustration (photograph or drawing) of a TYPE specimen.

ICZN *See* INTERNATIONAL COMMISSION ON ZOOLOGICAL NOMENCLATURE.

-IDAE A standardized suffix used to indicate an animal family in the acknowledged code of classification. The Varanidae, for example, is the family containing the monitor lizards.

IGUANIDAE Iguanas and chuckwallas. Family of the SAURIA, inhabiting North and South America including the West Indies, Galapagos Islands, Madagascar, Fiji and Tonga. Formerly with over 600 species in more than 50 genera, the family has recently been split into eight families: the CORYTOPHANIDAE, CROTAPHYTIDAE, HOPLOCER-CIDAE, IGUANIDAE, OPLUR-IDAE, PHRYNOSOMATIDAE, POLYCHROTIDAE, and the TROPIDURIDAE.

IHC Acronym denoting the International Herpetological Society (United Kingdom).

ILIUM (pl.) **ILIA** The largest, widest and uppermost of the three bones that constitute each half of the PELVIC GIRDLE in tetrapod vertebrates.

IMAGO An entomological term for an adult insect, but occasionally used in herpetology for the stage in the life cycle of an amphibian following METAMORPHOSIS.

IMBRICATE The term used to describe body scales that overlap in a way similar to the slates on a roof. *Contrast with* JUXTAPOSED.

imm. Abbreviation for immature.

IMMUNITY The ability of an organism to resist disease or toxins. The science concerned with the occurrence and causes of immunity is called immunology.

IMPACTION The condition in which hard faeces, undigested food items, or ingested stones, sand grains etc., have gathered and formed an impassable blockage in the alimentary canal.

IMPERVIOUS In reference to the inability of water to pass through or over the skin.

IMPREGNATION Fertilization; to cause to conceive; to fertilize via union of egg and sperm.

INAE A standardized suffix used to indicate an animal subfamily in the acknowledged code of classification.

INAPPROPRIATE NAME A name indicative of a particular feature, trait, quality, or ancestry which is not evident in the TAXON bearing that name.

INBREEDING Breeding between individuals that are closely related to each other, e.g., between parents and offspring or between SIBLINGS. *Contrast with* OUTBREEDING.

INCLUSION BODY DISEASE A highly transmittable and typically fatal disease affecting boas and pythons (BOIDAE), and commonly known as the boa bug. It is caused by a virus which, in its early stages, damages the spleen and nervous system eventually spreading to the kidneys and pancreas where it forms strange, dense microscopic bodies. Symptoms include the inability to constrict, elevation of the head vertically upward, and paralysis of the righting reflex.

INCOAGULABLE Said of blood that does not form a clot.

INCRASSATE Thickened or swollen.

INCUBATION The period of development of eggs, especially in an incubator, where heat and humidity are important factors

INCUBATION PERIOD The length of time necessary for an egg or eggs to develop, and varying from species to species. Generally expressed in days, the incubation period can be as short as 56 days in some snakes and lizards or as long as 340 days in certain chelonians.

INCUBATOR Any form of box-like apparatus in which reptile eggs can be artificially hatched under constant and sterile conditions.

INDIGENOUS Occurring, or living naturally, in a particular place or region, but not restricted to that place or region; a native. The European common frog *Rana temporaria*, for example, is indigenous to the British Isles, whereas the marsh frog (*Rana ridibunda*) is not, having been introduced there by man.

INDIVIDUAL DISTANCE The distance between individuals of the same species which when reduced incites aggression or avoidance.

IN-EGG In reptiles exhibiting OVIPARITY, said of a female carrying eggs.

INFRA- A prefix meaning below; beneath; under.

INFRACEPHALIC Pertaining to the VENTRAL, or undersurface, of the head.

INFRALABIAL The lower labial; any one of the enlarged LABIAL scales bordering the lip of the lower jaw.

INFRAMARGINAL A lamina situated between the PECTORAL and ABDOMINAL laminae of the plastron and the MARGINAL laminae of the carapace in most chelonians. *See* AXILLARY; INGUINAL.

INFRAMARGINAL GLAND In some chelonians, a musk gland situated at the point where the carapace joins the plastron

INFRAMAXILLARY An alternative term for the CHINSHIELD of snakes.

INFRAORDER The taxonomic category, used in the classification of organisms, that consists of one or more FAMILIES within an ORDER.

INFRA-RED Of, pertaining to, or made up of, radiation occurring within the part of the electromagnetic spectrum with a wavelength shorter than radio waves but longer than light. Situated below or beyond the red in the spectrum of ordinary light.

INFRASOUND In crocodilians, a form of acoustic communication caused by sub-audible body vibrations from rapidly contracted trunk muscles as the reptiles lie just below the water's surface. Barely perceptible by man, the sound waves radiating from a crocodilian's body have a very low frequency (1-10 hertz) and, underwater, can travel long distances. Used by a number of species during courtship, infrasound is frequently accompanied by a HEADSLAP and JAWCLAP.

INFRASPECIFIC NAME A broad term for any name below the rank of species. The term includes SUBSPECIFIC and INFRASUBSPECIFIC names.

INFRASUBSPECIFIC Used in reference to a group or name with a rank below that of a SUBSPECIES.

INFRATYMPANIC The term for any one of the scales in crocodilians, that occupy the lower margin of the opening to the ear.

INGUINAL ; INGUEN 1. Of, or relating to the groin; the area directly in front of the point where the hind limb joins the body. 2. In chelonians, the lamina on the rear border of the BRIDGE, between the CARAPACE and the PLASTRON.

INGUINAL AMPLEXUS A form of AMPLEXUS in which the males of certain anuran species grasp the females directly in front of the hind limbs, bending their bodies during spawning in order to bring their cloacae as close as possible to those of the females.

INGUINAL LAMINA In chelonians, any one of the INFRAMARGINAL laminae situated directly in front and to the side of the INGUINAL NOTCH.

INGUINAL NOTCH Either one of the two U-shaped notches situated at either side of the rear of the chelonian shell through which the hind limbs protrude.

INNOMINATE BONE The body of bone in reptiles that makes up each half of the pelvic girdle, resulting from the union of the ISCHIUM, ILIUM, and PUBIS.

INSECTIVORE An animal that feeds on insects. Many lizards and most anurans are insectivorous.

INSTINCT The innate ability of animals to behave in a particular genetically-fixed way in response to external stimuli. Instinctive reactions are performed involuntarily, e.g., an anuran's insect-catching tongue-flick is always executed in the same way whether prey is caught or not, and young marine turtles always make straight for the sea on hatching from their eggs even though they have never seen it before and may not be able to see it from the nest.

INSULAR Of, or relating to, islands.

INTEGUMENT One or more tissues forming the outermost body layer of an animal, serving to protect and insulate the body from its external environment.

INTEGUMENTAL POISON GLAND *See* NUCHAL VENOM GLAND.

INTER- A prefix meaning between; among.

INTERANAL A single LAMINA situated between the ANAL laminae of the PLASTRON and occurring exclusively in the marine turtles (CHELONIIDAE).

INTERCALARY Added; inserted, introduced.

INTERCALARY CARTILAGE In hylid frogs, an additional cartilaginous component situated between the penultimate and last bones of the digits.

INTERCALARY REPLACEMENT; ALTERNATE REPLACEMENT Tooth replacement, characteristic of amphisbaenids and certain lizards, in which new teeth lie at an angle at the base of, and in between, the functioning teeth, moving obliquely into empty sockets as and when the others are lost.

INTERCANTHAL In rattlesnakes, any one of the many small scales on top of the snout between the CANTHAL scales where, in other snake species, the larger, regular head scales, or plates, are situated.

INTERCHINSHIELD Any one of the scales situated between the elongated and paired CHINSHIELD scales in certain snakes, e.g., several rattlesnake species.

INTERDIGITAL MEMBRANE The WEB that joins the digits of many amphibians and reptiles.

INTERFEMORAL Any one of the scales occupying the area between the hind legs on the under surface of lizards.

INTERFEMORAL SEAM The narrow line running between the FEMORAL laminae in the chelonian PLASTRON

INTERGENIAL In snakes, any one of the scales situated between the large, paired CHINSHIELDS.

INTERGENUAL EXTENT In anurans, the measured space between the knees when the FEMORA are extended at right angles to the body.

INTERGRADE Crossbreed or HYBRID; the result of interbreeding of closely related species or subspecies in the wild (e.g., *Vipera aspis* x *Vipera berus*) or in captivity (e.g., *Python sebae* x *Python molurus bivittatus*).

INTERGULAR A single lamina situated between the GULAR laminae of the plastron in the suborder Pleurodira (side-neck turtles).

INTERHUMERAL SEAM The narrow line running between the HUMERAL laminae in the chelonian plastron.

INTERNAL FERTILIZATION The vital process of sexual reproduction occurring within the female's body and usually an adaptation to life in a terrestrial environment. Internal FERTILIZATION occurs in all reptiles, and in a few amphibians such as *Ascaphus, Nectophrynoides* and *Eleutherodactylus*, for example. *Contrast* EXTERNAL FERTILIZATION.

INTERNAL GILL The organ of respiration, developed in most fishes, formed as outgrowths from the wall of the pharynx and situated in gill slits. It is ventilated as water, drawn in through the mouth, is forced from the pharynx, past the gills and out through the gill slits. The larvae of amphibians have EXTERNAL GILLS which project from the body so that water passes over them as the animal moves.

INTERNAL NARES *See* NARIS.

INTERNARIAL DISTANCE The measured space between the NARES, or nostrils, the measurement of which can aid identification in some species.

INTERNASAL; ANTERIOR FRONTAL Any one of the enlarged scales, or plates, in lizards and snakes on the dorsum of the head at the front of the snout between the NASAL and directly behind the ROSTRAL scales. Usually paired in snakes.

INTERNASAL RIDGE The sharp rim on the tip of the snout and CANTHAL RIDGE in *Crotalus willardi* shaped by the upward turn of the outer edges of the INTERNASAL and CANTHAL scales.

INTERNASAL SPACE In amphibians, the area of skin on the head between the external nostrils.

INTERNATIONAL CODE OF ZOOLOGICAL NOMENCLATURE The system of rules, regulations, and recommendations governing the constancy and universality in the scientific naming of animals, making certain that each taxon's name is both different and exclusive.

INTERNATIONAL COMMISSION ON ZOOLOGICAL NOMENCLATURE; ICZN The international organization that drafts and manages the use of the rules, regulations, and recommendations of the INTERNATIONAL CODE OF ZOOLOGICAL NOMENCLATURE.

INTERNATIONAL ZOOLOGICAL YEARBOOK An annual publication, published by the Zoological Society of London, intended for keepers and workers in zoos and wildlife parks around the world. Each book is separated into three chief sections. Section 1 deals with the special topic of that particular year. Section 2 reports on new developments in the zoo world, e.g., breeding successes and husbandry techniques, research, building construction, and exhibits. Section 3 is a reference section containing information on the world's zoos, aquaria and bird parks together with breeding lists and rare animal cencus, sudbooks and registers etc.

INTEROCCIPITAL 1. In crocodiles a small, centrally situated scale on the rear of the head between the OCCIPITAL scales. **2.** In certain lizards, a centrally situated enlarged scale, or plate, on the back of the head, situated directly behind the INTERPARIETAL.

INTEROCUFRONTAL In the VIPERIDAE, any one of the scales on the head that occupy the area between the FRONTAL and SUPRAOCULAR scales.

INTEROCULABIAL Any one of the scales situated below the eye and between the SUBOCULAR and SUPRALABIAL scales in some snake species, e.g., rattlesnakes.

INTEROCULAR Used in reference to the area on top of the head between the outer margins of the eyelids. For example, any pale or dark narrow stripe on the top of the head between the eyes is termed an 'interocular bar'.

INTEROCULAR BAR *See* INTEROCULAR.

INTEROCULAR WIDTH In snakes, the measured space between the nearest sites of the border of the eye and the midline of the head

INTERORBITAL In certain snakes and lizards, any one of the small, irregular scales occupying the area, between the ORBITS, on the back of the head, as seen in many boids for example.

INTERORBITAL DISTANCE **1.** In anurans, the space between the ORBITS calculated over the least distance, typically the midpoints of the eyes. **2.** The distance between the front corners of the eyes.

INTERPARIETAL In many lizards, a median scale situated between the PARIETAL scales on the top of the head and the site of the external opening of the PARIETAL EYE in some species.

INTERPECTORAL SEAM The narrow line running between the PECTORAL laminae in the chelonian PLASTRON.

INTERPLASTRAL In certain chelonians e.g, *Chelydra*, a LAMINA on the middle ventral line of the PLASTRON.

INTERPREOCULAR In rattlesnakes, any small scale inserted between the upper and lower PREOCULAR scales.

INTERRUGAL SPACE In lizards of the genus *Anolis*, the slightly sunken surface occupying the space between the ridges, or rugae, on the top of the head.

INTERSCALAR SKIN; INTERSTITIAL SKIN Of, or pertaining to, the usually thin skin lying between the scales of reptiles

INTERSPACE The area of pigmentation separating one pattern element from another on the backs of snakes and lizards.

INTERSPECIFIC Between two or more species. *Compare* INTRASPECIFIC.

INTERSTITIAL Occurring in the spaces between tissues, organs, etc.

INTERSTITIAL SKIN *See* INTERSCALAR SKIN.

INTERSUPRAOCULAR Any one of the often irregularly placed scales on top of the head, situated between the SUPRAOCULAR scales.

INTERXIPHIPLASTRAL NOTCH In chelonians, the notch situated posterior to the point where the paired XIPHIPLASTRON plates meet on the anal part of the PLASTRON.

INTRA- A prefix meaning inside; within.

INTRAMUSCULAR Within the muscles.

INTRASPECIFIC Within a single species. *Compare* INTERSPECIFIC.

INTRAVASCULA Within the blood vessels.

INTRAVENOUS Within the veins.

INTRO- A prefix meaning inward.

INTRODUCTION Referring here to any individual animal or species brought from an area where it occurs naturally to an area where it is not INDIGENOUS; also termed 'naturalization'. Such introductions are termed 'aliens' or 'exotics', e.g., the cane toad *Bufo marinus* is an alien to Australia where it has been introduced by man.

INTROGRESSION The introduction of genes from one species into the GENE POOL of another by HYBRIDIZATION.

INTROMISSION The alignment of the cloacae and insertion of the male's HEMIPENIS (in snakes and lizards) or PENIS (in chelonians, crocodilians and caecilians) into the VAGINAL POUCH of the female.

INTROMITTENT ORGAN The male copulatory organ in animals which reproduce by INTERNAL FERTILIZATION, e.g., the HEMIPENIS of lizards and snakes, the PENIS of crocodilians, caecilians and chelonians, and the TAIL in frogs of the ASCAPHIDAE.

INVALID NAME In nomenclature, any AVAILABLE or UNAVAILABLE NAME, but which is not the VALID NAME, allocated to a TAXON.

INVERTEBRATE Any animal lacking a vertebral column, and including such diverse forms as worms, ARTHROPODS (including insects), and molluscs. The diets of most amphibians and many lizard species consist almost entirely of invertebrates.

INVOLUTION The regression and eventual disappearance of an organ.

IN YOUNG In reptiles exhibiting OVOVIVIPARITY or VIVIPARITY, said of a female carrying unborn young

IONIDES, Constantine Philip, (1901-1968), naturalist, game-ranger and herpetologist with a worldwide reputation. During his numerous safaris, in which he visited every country of Central Africa, he contributed greatly to herpetological knowledge, discovering several new forms and devising new methods of catching and handling the more dangerous species. In addition he played a very important role in the development of the National Museum in Nairobi, personally collecting many of its rarer animals.

IRIDESCENCE The rainbow lustre in reptiles, resulting from the arrangement and structure of the scales and occurring usually as a purple, green or bluish sheen which changes colour with a change in the angle of view. Iridescence is particularly noticeable in the sunbeam snake *Xenopeltis unicolor*. In many species of snakes, certain lizards and several amphibians, it is considered by herpetologists to be a sign of good health.

IRIDOPHORE A type of CHROMATOPHORE containing purines which, unlike the MELANOPHORE and xanthophore, does not produce colours but diffracts light, which results in various iridescent effects, such as the green colour of many anurans which takes on a bluish hue when seen in a different light.

IRIS The often brightly coloured, circular area of muscular tissue in the front of the eye, serving as a diaphragm to control the size of the PUPIL and thereby regulating the amount of light entering the eye.

ISCHIUM The rearmost of the three bones that constitute each half of the pelvic girdle in tetrapod vertebrates.

ISCHAEMIA Restricted and brief anaemia resulting from an obstruction of blood circulation to a body part.

ISIS Acronym denoting the International Species Inventory System. ISIS assembles, administers and disseminates biological information on animal species held in captivity. With the help of ISIS many endangered animals are now being successfully bred in captivity.

ISODONT Having teeth that are comparatively regular in both appearance and size, as in most lizards and non-venomous snakes.

IUCN Acronym denoting the International Union for the Conservation of Nature (and natural resources). The IUCN/World Conservation Union is the only truly worldwide conservation organization with more than 10,000 internationally recognized scientists and experts in some 140 countries. Established in 1948 as the International Union for the Protection of Nature (IUPN), it became the International Union for Conservation of Nature and Natural Resources in 1956 and in 1990 shortened its name to IUCN - The World Conservation Union. The Union links together people and both government and non-government institutions and encourages a worldwide approach to conservation by managing and restoring ecosystems and protecting threatened species.

IUPN *See* IUCN.

IXODIDAE *See* TICK.

IXODIDES *See* TICK.

IZY Abbreviation for INTERNATIONAL ZOOLOGICAL YEARBOOK.

J

JACARE; YACARE A Brazilian name for the CAIMAN. Examples include jacare-tinga, the spectacled caiman *Caiman crocodilus*; jacare-curua, the broad-snouted caiman *Caiman latirostris*; and jacare-acu, the black caiman *Melanosuchus niger*. **JACKSON'S METHOD** An old method of treatment once used for the bites of VENOMOUS snakes. Developed by Dr Dudley Jackson, it involved cutting across and around the fang punctures with cross-shaped incisions. As the swelling at the site of the bite increased, so the incisions were lengthened, whilst suction was applied to them.

JACKSON'S RATIO Formulated by Dr Oliphant Jackson, an eminent veterinary surgeon who did much important work on reptile health and husbandry. His ratio is one of the criteria that can be used to show the state of health of spur-thighed and Hermann's tortoises (*Testudo graeca* and *Testudo hermanni*) when they are maintained in captivity. By comparing the length of the CARAPACE with the weight of a particular tortoise, the resulting point, plotted on a graph, can indicate whether or not that tortoise is in good health.

JACOBSON'S ORGAN; VOMERONASAL ORGAN A modified part of the nasal sac forming the major sense organ in many reptiles, it consists of a pair of pouch-like structures embedded in the frontal region of the palate and lined with sensory tissue. In an active snake, the tongue is constantly protruded and withdrawn. Upon withdrawal, the forked tips are thrust into these spiral, blind-sac openings and any particles that have been picked up from surrounding objects or the air are conveyed to the sensory lining of Jacobson's organ for analysis by the brain, via a special branch of the OLFACTORY nerve. The organ is absent in adult crocodilians.

JACURA The northern tegu *Tupinambis nigropunctatus* of the TEIIDAE family, inhabiting central and northern South America.

JAN, George, (1791-1866), Italian zoologist. Compiled *Iconographie generale des ophidiens*, a three-volume work, illustrated by Sordelli, describing all the snakes then known.

JARARACA *Bothrops jararaca*, a PIT VIPER of the family VIPERIDAE, subfamily Crotalinae, inhabiting Brazil, Paraguay and northern Argentina.

JARARACUSSU *Bothrops jararacussu*, a PIT VIPER of the family VIPERIDAE, subfamily Crotalinae, inhabiting Brazil, Paraguay and northern Argentina.

JAUNDICE A condition affecting tortoises and usually a result of either dehydration or fatty degeneration of the liver due to incorrect diet. Affected individuals are reluctant to feed and inclined to hide in corners or bury themselves.

JAWCLAP An acoustic and subsonic signal employed by most crocodilian species especially during courtship. The opened jaws are suddenly and sharply clapped shut either on or below the water surface, an action frequently accompanied by a HEADSLAP. The jawclap appears to act as a deterrent to rival males and as an attraction to females, which sometimes respond with their own jawclap.

JELLY Gelatinous material exuded partly from glands in the walls of the oviducts and partly from glands in the CLOACA in amphibians. Affixed to the eggs as they pass through, the jelly may be set down in concentric layers, or as jelly envelopes, the number of which varies with the species, or it may form an uninterrupted tube without any noticeable layering, as in the JELLY STRAND of members of the BUFONIDAE.

JELLY ENVELOPE; EGG CAPSULE In the amphibian egg, any one of the several concentric layers of the translucent viscous substance deposited on it by glands in both the OVIDUCT and CLOACA as it passes down the oviduct.

JELLY FLOAT In eggs of the frog *Kaloula*, a wide, marginal structure, formed soon after ovulation in the exterior jelly envelope of each egg, enabling the CLUTCH to float at the water's surface.

JELLY STRAND The often considerably lengthy string of JELLY that coate the eggs of toads of the genus *Bufo*. Bufonid jelly is made up of four layers: the two central ones envelop each individual egg and the two outer ones are common to the whole CLUTCH.

JIZZ Originally a term used by birdwatchers which has recently found its way into herpetology; the combination of delicate and indefinable characteristics that is frequently the simplest method of recognizing and isolating species that are closely related. The successful use of jizz relies largely upon the observer having some experience of the animals concerned. *See* GESTALT.

JORDAN'S RULE The closest-related individuals of a particular species do not occur in the same region, or in a distant one, but in an adjoining area, either partly or completely separated by an obstacle or boundary.

JUBAL In lizards of the genus *Eumeces*, the greatly enlarged scale, or scales, located directly behind the head.

JUGAL 1. Used by some authors to describe a scale situated below the eye in crocodiles, with 'first jugals' being the row of scales on the margin of the lower eyelids and 'second jugals' the row below those. **2.** In reptiles and certain amphibians, particularly caecilians, a dermal bone in the BUCCAL area, ventrally bordering the eye and extending behind the upper jaw bone (MAXILLA) forming part of the lipline.

JUGULAR Of, or relating to, the neck and/or throat. In snakes, any one of the scales situated between the CHINSHIELD and lower LABIAL scales.

JUNGLE CHICKEN The colloquial name for the Goliath frog *Conraua goliath* of West Africa, the world's largest frog exceeding 300 mm in length.

JURASSIC The middle period of the MESOZOIC era, from 195 million to 135 million years ago. Reptiles were large and plentiful, dominating the vertebrates, and the pterosaurs, the first flying reptiles, occurred. Fossils of the earliest-known primitive bird (*Archaeopteryx*) and of the first mammals are found in Jurassic rocks

juv Abbreviation for JUVENILE.

JUVENILE A very young, newly hatched or newborn individual, often displaying proportions and colorations differing from those of the ADULT.

JUXTAPOSED Term used to describe body scales that do not overlap but are disposed in a side-by-side fashion. *Compare* IMBRICATE.

K

KARUNG In the commercial skin trade, the term for a snake skin used in the manufacture of clothes, shoes and bags etc., in which the scales are crushed to create a skin of uniform appearance. Certain species, notably the wart snake *Acrochordus javanicus*, are particularly valuable in this respect.

KARYOTYPING Characteristics of cell nuclei, in particular size, form and number of chromosomes.

KEEL 1. Any ridge running from front to back on the CARAPACE or PLASTRON of a chelonian.
2. The slightly raised line, or ridge, on the middle of a single scale in some species of snakes, the distribution, presence (or absence) and definition of which are all important in CLASSIFICATION.
3. The elevated border along the upper surface of the tail in some urodeles, which can sometimes extend on to the body. The dorsal fin on anuran tadpoles is also termed a keel.

KERATIN A tough, fibrous, sulphur-containing protein that occurs in the outer layer (EPIDERMIS) of the skin of both amphibians and reptiles, and which furnishes the rudimentary constituent for the formation of scales, spines, claws and horns etc.

KERATINOUS; CORNIFIED Having a thick protein that forms the external layer of the skin in many vertebrates and in reptiles the scales, claws and nails.

KERATITIS Inflammation of the cornea; most common in chelonian genera.

KERATOPHAGY The eating of KERATIN or keratinous materials. The majority of amphibians and some reptiles (mainly lizards of the family GEKKONIDAE) devour all or part of their skins during the act of sloughing, or ECDYSIS. The dead, cornified cells in the sloughed EPIDERMIS are an important supply of protein too valuable to waste.

KEY A systematic list or table of diagnostic, taxonomic key characteristics assisting in the swift identification of animals and plants.

KEY CHARACTERISTIC *See* KEY.

KIDNEY DISEASE Any one of several infectious (NEPHRITIS) or non-infectious (NEPHROSIS) diseases of the urinary organs, of which GOUT is the most significant in reptiles. Causes of kidney disease in reptiles include vitamin deficiency, parasite infestations and *Pseudomonas* infections.

KINASE Enzymes that catalyse the transfer of energy groups between cells.

KINETIC Pertaining to, characterized by, or caused by motion.

KINETICISM; KINESIS Motion; normally used in herpetology in reference to movement inside the skull. Chelonian and crocodilian skulls are AKINETIC whereas lizard

112

and snake skulls normally display much mobility.

KINGDOM The highest ranking category used in the taxonomic classification of organisms.

KINK The sharp crook occurring in a digit of the frog genus *Rhacophorus,* brought about by an extra cartilaginous component, the INTERCALARY CARTILAGE.

KINOSTERNIDAE Mud and musk turtles. Family of the CHELONIA, suborder Cryptodira, inhabiting eastern Canada and extending into South America. Approximately 20 species in two genera.

KLAUBER, Laurence Monroe, (1883-1968), American herpetologist who carried out intensive and widespread studies of rattlesnakes (CROTALIDAE), culminating in the two-volume *Rattlesnakes; their habits, life histories and influences on mankind* (1956). He also produced many papers on other reptile taxa of the south-western USA.

KNOB 1. A rounded projection, lump, stud or boss and used in reference to the enlarged keel on scales located near the anus in some species of snakes, termed 'knobbed anal keels'. **2.** A collective noun used sometimes for a gathering of toads.

KNOBBED ANAL KEEL *See* KNOB.

KNOT A collective noun for a gathering of spawning frogs or toads, less commonly for a mating snakes.

KRAIT A venomous Asian snake species of the genus *Bungarus,* family ELAPIDAE, inhabiting

cultivated areas, meadows, and wooded plains, often adjacent to rice paddies, streams, rivers and ditches, in parts of India, Pakistan, China and the Malay Peninsula, Java, Borneo and Sumatra.

KRONISM The killing and devouring of offspring.

KUFI Levantine or blunt-nosed viper *Vipera lebetina.* A member of the VIPERIDAE family of the SQUAMATA, subfamily Viperinae, inhabiting Middle East, north Africa, north-west Pakistan and the Cyclades. A bold and dangerous species attaining 120 centimetres in length.

KYPHOSIS Hunchback; the backward curvature of the spine resulting either from injury or disease, or may be CONGENITAL in origin. Occurs predominately in chelonians where it manifests itself as an abnormality in the CARAPACE.

L

L *See* LINNAEUS.

LABIA (sing.) **LABIUM** A lip or liplike structure.

LABIAL Of, or pertaining to, the upper or lower lip (labium) in reptiles; any one of the row of scales bordering the mouths of snakes and lizards on the upper and lower lips and termed upper labial and lower labial respectively.

LABIAL GLAND In the genus *Salamandra,* a glandulous area in the skin of the lips and chin. **LABIA PAPILLAE** *See* ORAL PAPILLAE.

LABIAL PIT Any one of a number of cavities, varying in size and disposition from one genus to another, each set in an individual upper or lower LABIAL scale; found in many snakes of the BOIDAE family and in one or two other species, and serving as highly sensitive thermoceptors able to detect temperature differences as low as 0.025°C. The labial pit also enables the snake to seek and find warm-blooded prey in complete darkness and is similar in both structure and function to the PIT of the Crotalinae, or PIT VIPER group.

LABIAL TEETH Small, horny cusps situated, in transverse rows like the rasps on a file, on the lips (LABIA) of anuran larvae.

LABIUM (pl.) **LABIA** A lip, or any lip-like structure.

LACÉPÈDE, Bernard, (1756-1825), French zoologist. Described many new reptile and amphibian species, and his four-volume *Natural history of oviparous quadrupeds and serpents* (1789), was the first comprehensive world summary of reptiles and amphibians.

LACERTID Any lizard of the family LACERTIDAE.

LACERTIDAE 'Typical' lizards. Family of the SQUAMATA, suborder SAURIA, inhabiting most of Europe, Asia and Africa (except Madagascar). Over 200 species in some 20 genera.

LACERTILIA An alternative term, preferred by some authors, for the SAURIA (lizards), the suborder of SQUAMATA.

LACHRYMAL PIT *See* PIT

LACRIMAL In rattlesnakes, the enlarged scale situated between the PREOCULAR and POSTOCULAR scales and bordering the lower frontal area of the ORBIT

LACRIMAL GLAND In reptiles, one of the two large glands connected with each eye, that produce the 'tears' that help to keep the eye moist. It is situated towards the back of the eye, its secretions emptying into the conjunctival space under the eyelid where they lubricate and cleanse the exposed eye surface. Closely associated with the NICTITATING MEMBRANE of chelonians, crocodilians and lizards. *See* HARDERIAN GLAND.

LACRYMAL An alternative spelling for LACRIMAL, used in earlier literature.

LACUNAL Any one of the enlarged and outward curving scales forming the inner, and part of the outer, borders of the PIT in crotaline snakes.

LACUNOLABIAL A large, single scale in certain crotaline snakes, resulting from the union of the prelacunal and the second (or occasionally , the third) SUPRALABIAL scales.

LAEVOGYRINID A term given to most anuran larvae which possess a single SPIRACLE on the left side of their bodies.

LAMELLA (pl.) **LAMELLAE** In herpetology, used most frequently for the series of thin, either single or divided, transverse plates extending across the underside of the digits in many lizard species.

LAMINA (pl.) **LAMINAE** Any one of the horny, epidermal scutes covering the bony plates in the skeleton of the CARAPACE and PLASTRON of many chelonians. The term is used less frequently today, many authors preferring the word 'scute'.

LAMINAR NUCLEUS The slightly off-centre, JUVENILE area (AREOLA) of the lamina in many chelonians.

LAMINATED Arranged in a succession of slender, parallel plates or sheets.

LAND BRIDGE A link between two land masses, particularly continents, that permits animal (and plant) migration between them. Prior to the general recognition of continental drift, the occurrence of past land bridges was frequently cited to explain similarities in the fauna and flora of continents that are now widely separated.

LANTHANOTIDAE The earless monitor *Lanthonotus borneensis*. Family of the SQUAMATA, suborder SAURIA, inhabiting Borneo. MONOTYPIC.

LARGE-SCALE HIDE A term used in the commercial skin trade for soft-bellied species of crocodilians that have 20-25 transverse ventral scale rows. *See* SMALL-SCALE HIDE; TRANSVERSE VENTRAL SCALE ROWS.

LARVA (pl.) **LARVAE** The stage in the life of many animals, e.g., many invertebrates, most amphibians and certain fishes, that follows the embryo and precedes the adult stage. A larva very often bears little or no resemblance to its parents and must undergo a METAMORPHOSIS before it assumes the characteristic features of the adult.

LARYNX The upper portion of the TRACHEA (windpipe) of tetrapod vertebrates which, in reptiles and amphibians (and mammals), contains the vocal cords.

LATASTE, Fernand, (1847-1934), French herpetologist and professor of medicine. His work in herpetology was involved principally with the amphibians of north-western and western Europe.

LATERAL Pertaining to, or situated on, the sides; any one of the scales on the side of a lizard's body that are neither DORSAL nor VENTRAL scales.

LATERAL DERMAL FOLD An alternative term for the DORSOLATERAL FOLD on the sides of the body in some amphibians and occurring in a number of lizard species also. Frequently shortened to dermal fold.

LATERAL LAMINA An alternative term for the COSTAL lamina in many chelonians.

LATERAL LINE SYSTEM A series of sense organs (NEUROMASTS) disposed on the head and along the body in aquatic amphibians, responsible for the detection of sounds and changes in pressure, in water.

LATERAL MARGINAL An alternative term for the MARGINAL laminae of chelonians.

LATERAL UNDULATION *See* SERPENTINE LOCOMOTION.

LD *See* LETHAL DOSE.

LD$_{50}$ The median lethal dose (LD) of a substance that will kill half (50%) of a group of experimental animals to which it is exposed. It is frequently used as a standard measure of snake-venom toxicity.

LEGUAAN The South African name (likkewaan in Afrikaans) for a monitor lizard or goanna and derived from the name 'iguana'.

LEPIDOSIS; SCALATION Pertaining to the number, pattern, shape, position and arrangement of scales in reptiles.

LETHAL DOSE The amount of VENOM required to cause the death in a test animal.

LEECH Any freshwater or terrestrial annelid worm of the class Hirudinae. Leeches are either carnivorous predators or blood-sucking parasites, feeding upon the blood or tissues of other animals. Chelonians and crocodilians are frequently parasitized by leeches of the genus *Haementeria*, due mainly to the fact that both the leeches and the reptiles lead aquatic or semi-aquatic existences and share the same habitats.

LEIOPELMATIDAE Family of the ANURA, inhabiting the north-western USA and New Zealand. Four species in two genera.

LENTIC Of, inhabiting, or relating to, still waters.

LENTICULAR Shaped like a biconvex lens.

LEPIDOSAURIA A subclass of the more primitive DIAPSID reptiles that includes the tuatara (RHYNCHOCEPHALIA) and the snakes and lizards (SQUAMATA).

LEPIDOSIS The pattern and disposition of scales.

LEPTODACTYLIDAE Family of the ANURA, inhabiting Central and South America. Over 700 species in 52 genera.

LEPTOTYPHLOPIDAE Thread snakes. Family of the SQUAMATA, suborder OPHIDIA, inhabiting tropical America, Africa and extending into western Asia. 50 species in two genera.

LETHAL DOSE *See* LD$_{50}$.

LETISIMULATION *See* DEATH FEINT; THANATOSIS.

LEUCISTIC *See* LEUCISTISM.

LEUCISTISM A condition in which the body of an animal is partially or completely white, or colourless, but with eyes that retain their usual pigmentation (eye pigment is lacking in a true ALBINO). Such leucistic animals are sometimes termed 'dilute albinos'.

LID OEDEMA A disease of the HARDERIAN GLAND, usually associated with an abnormal accumulation of fluid causing swelling and resulting in protrusion of the eyelids, occurring in a number of chelonian genera.

LIGHT ADAPTED Descriptive of anurans housed in the laboratory in bright light. Frogs maintained in such conditions for some hours allow experiments to be undertaken to ascertain the effect of light on skin pigmentation. *See* DARK ADAPTED.

LIMB BUD STAGE The phase in the life of larval anurans when the limbs are nothing more than plain bud-like outgrowths lacking any recognizable external features.

LIMITS OF TOLERANCE The upper and lower limits to the range of specific ecological aspects (e.g., temperature, light, water availability) within which an organism can continue to exist. Reptiles and amphibians with a broad range of tolerances are typically widely distributed, whilst those with a limited range have a more restricted distribution. *See* SHELFORD'S LAW OF TOLERANCE.

LIMNOLOGY The study of the flora, fauna, and ecology of river systems.

LINEAGE *See* EVOLUTIONARY LINEAGE.

LINE TRANSECT A sampling method used to measure the distribution of one or more species in a given habitat in which a rope, string or tape is stretched out on the ground in a straight line between two posts. Only species actually touching the line are taken into account in the sampling. *Compare* BELT TRANSECT.

LINGUAL Of, or relating to, the tongue.

LINGUAL FOSSA In snakes, the small V-shaped indentation in the underside of the ROSTRAL scale, through which the tongue can protrude while the mouth remains shut.

LINGUATULID A modified, worm-like ARACHNID, occurring almost exclusively in reptiles as an ENDOPARASITE.

LINN *See* LINNAEUS.

LINNAEUS; L; LINN; VON LINNÉ, Carl, (1707-1778), Swedish naturalist, botanist and physician. Established the Stockholm Academy of Sciences and exerted an influence in his fields of natural history and botany that has had few parallels in the history of science. Famous for formulating the basic principles for classification, and for founding modern systematic botany in 1753. Three years later he presented, for the first time, in the tenth edition of his work, *Systema Naturae,* a binomial (two-name) system of NOMENCLATURE, in which each species received two names: the first one for the genus; the second for the particular species. He immediately received acclaim from scientists throughout the world and the system, named after him, is in universal use today. In 1788, the Linnéan Society was founded in England in his honour, publishing transactions and journals on all matters of natural history.

LIPOPHORE A pigment cell of the skin associated with red and yellow coloration.

LIQUID INTAKE Most reptiles satisfy their liquid requirements by drinking fresh water. Snakes and chelonians immerse the tips of their snouts in water and swallow using muscular movements of their throats, and lizards lap up water droplets with their tongues. Many desert species rarely drink water, receiving most, if not all, of the liquid they require from the food they eat. In aquatic amphibians, liquid absorption takes place through the whole surface of the body, including the cloaca, while they are submerged in water; other species obtain any moisture they may need from their prey.

LITTER 1. The offspring resulting from a multiple birth. **2.** The detritus of fallen leaves and bark which accumulates over a period of time on the floor of woods and forests etc.

LITTER LAYER The deposit of organic material that carpets the surface of the soil.

LITTORAL Of, relating to, or inhabiting, the shallow waters or shores of a sea, ocean (marine littoral) or lake (freshwater littoral).

LIZARDLET A rather inelegant term used very occasionally for a newborn or hatchling lizard.

LIZARD POX *See* POX.

LOBE Any rounded or curved projection that forms part of a larger structure. Used in herpetology for: **1.** The area on a chelonian PLASTRON directly in front and to the rear of the BRIDGE and termed 'anterior lobe' and 'posterior lobe' respectively. **2.** Any one of the enlarged areas on any one of the horny rings that constitute the rattle in *Crotalus* and *Sistrurus*.

LOCOMOTION The power, ability, or act of moving from one place to another. Locomotion in reptiles and amphibians takes many forms, from the lateral movements of the body and tail in aquatic urodeles and amphibian larvae to the trot and gallop in certain lizards and crocodilians. Limb reduction, present in varying degrees in many lizard families, amphisbaenids, caecilians, several urodeles and, ultimately, snakes, has resulted in the development of particular forms of locomotion to suit their different modes of life and adapted to the variable habitats in which they live. These include CONCERTINA, RECTILINEAR, SERPENTINE and SIDEWINDING MOVEMENT.

LONGEVITY Literally, 'long life', but used generally for the period in an organism's life from its hatching or birth to its natural death. The longevity, or life span, of reptiles and amphibians can vary from as little as one or two years in some of the smaller species to well over 150 years in larger species, e.g., the Aldabran giant tortoise *Geochelone gigantea*.

LONGITUDINAL Of, or relating to, length; located or running along the length.

LONGITUDINAL TYPE; RECTIFORM Used in reference to a particular arrangement, common to most snakes, in the scalation. *See* SCALE ROW.

LORA The arboreal parrot snake *Leptophis ahaetulla* of the family COLUBRIDAE, subfamily Colubrinae, inhabiting the dry forest areas of Mexico through to Argentina.

LORAL An old, seldom used alternative spelling for LOREAL.

LORE The space, usually in the form of a very shallow depression, on the surface of the head of snakes directly in front of the eye.

LOREAL A scale situated between those of the nostril (NASAL scales) and those of the eye (PREOCULAR scales), but not touching either, in snakes which usually have just one, and in lizards which may have several.

LOREAL PIT; FACIAL PIT The PIT on the side of the head of crotaline snakes (pit vipers).

LOREAL SCALE The scale situated between those of the nostril (NASAL SCALES) and those of the eye (PREOCULAR SCALES) but not touching either, in most snakes, which usually have just one, and in

lizards, which may have several. This scale is the singular most important scale for determining whether a COLUBRINE snake is a front-fanged elapid, or rear-fanged, or a non-venomous colubrid. All elapids (with the exception of some W. African populations of tree cobra *Pseudohaje goldii*) lack a loreal scale while most colubrids (exceptions include the white-bellied mangrove snake *Fordonia leucobalia*) possess a single loreal scale. A few species, especially boids, possess many fragmented loreal scales.

LORICA (pl.) **LORICAE** A hard outer covering, or CUIRASS; used in herpetology for the protective, bony armour of plates, shields and laminae of crocodilians and chelonians.

LORICATE Having a hard protective covering, or CUIRASS.

LORILABIAL In lizards, any one of a number of scales situated in a longitudinal row between the LOREAL and SUPRALABIAL scales.

LOTIC Descriptive of habitats that are created in running water (i.e., streams and rivers).

LOWER LABIAL *See* LABIAL.

LOWER ROSTRAL The scale situated at the centre of the tip of the lower jaw in snakes and lizards, bordered on either side by the first lower labials.

LUMBAR In crocodilians and lizards etc., the area of the back directly in front of the hind limbs.

LUMBAR GLAND In certain anurans, e.g., the leptodactylid *Pleurodema bufonia* and allied species, a distinct, enlarged and glandulous area of the lower back and sides between the pelvis and the abdomen.

LUMEN (pl.) **LUMINA** The space, cavity or canal enclosed within a duct, tube or similarly-shaped organ, of which the central passage of the digestive tract is an example.

LUMEN EXIT In venomous snakes, the point where the VENOM DUCT opens at the base of the FANG, allowing venom to flow down or through it.

LUNATE Crescent moon-shaped.

LUNGWORM Any one of a number of parasitic NEMATODE worms, especially of the genus *Rhabdias,* occurring in the lungs of many amphibians, snakes and some lizards. Untreated infestations in captive animals often result in death.

LURE 1. The bright, contrastingly-coloured tip of the tail in a number of juvenile snakes, such as certain arboreal species of *Trimeresurus*, that can be elevated and waved around, thereby attracting inquisitive prey to approach within striking distance. **2**. The flesh-coloured, motile worm-like projection from the tongue of the alligator snapping turtle (*Macroclemys temmincki*) that serves to attract small fish as the reptile lies submerged with its mouth agape.

LYMPH A colourless alkaline liquid, consisting mainly of white blood cells, collected from the tissues and conveyed around the body by the lymphatic system.

LYMPHATIC SYSTEM An broad network of capillary vessels that transports the INTERSTITIAL fluid of the body as LYMPH to the venous blood circulation.

LYMPHADENITIS Inflammation of one or more of the lymphatic vessels (nodes).

LYMPHADENOPATHY Enlargement of the LYMPH NODES which may be sore to the touch if severely inflamed.

LYMPH NODE Any one of many gland-like structures occurring along the course of the lymphatic system that help protect the body against infection by neutralizing toxins and killing bacteria.

LYOPHILIZATION *See* FREEZE DRYING.

LYSIN Any of a group of agents, such as antibodies for example, that cause the break up and dispersion of cells against which they have been directed.

LYSIS (pl.) **LYSES** The destruction of cells by the action of a particular LYSIN.

M

m Abbreviation for male.

MACRO- A prefix meaning large, or long; great.

MACROCEPHALY The condition of having a grossly enlarged head or skull, occurring occasionally in certain aged individuals of some Chelonians.

MAJA The Cuban boa *Epicrates angulifer.*

MALACHITE GREEN A substance often used in the treatment of various fungal infections of the shells of captive aquatic chelonians.

MALADAPTATION SYNDROME Bad or deteriorating health in captive animals as a result of inadequate care and bad nutrition.

MALAR Of, or relating to, or on, the cheek or cheekbone.

MALAR SCALE An enlarged scale on the lower jaw of *Amphisbaena.*

MALE RELEASE CALL In anurans, a vocal sound uttered by either or both male and female during AMPLEXUS, if and when grasped by one or more other males. The call informs other males that a pairing has already taken place and causes them to release their grasp.

MALPIGHIAN LAYER The innermost layer of dividing cells at the base of the epidermis of vertebrates. The cells contain MELANIN and move gradually upwards through the epidermal layers where they become hardened, forming the STRATUM CORNEUM, replacing those cells which are constantly being sloughed off.

MAMBA An aggressive, dangerous and venomous African snake of the genus *Dendroaspis*, family ELAPIDAE. There are four species including the black mamba *Dendroaspis, polylepis* and the green mamba *Dendroaspis, angusticeps*

MAMBIN; DENDROASPIN A potent component of the venom of African mambas (*Dendoaspis* sp.). Like DISINTEGRIN, it act as an antagonist inhibiting the blood from clotting.

MAMUSHI A Japanese name for the Pallas' pit viper *Gloydius (Agkistrodon) halys halys*, which is responsible for many incidents of snake bite in Japan each year, although fatalities are comparatively rare. The name is also often used for *Gloydius blomhoffi.*

MANDIBLE The cartilage and bone (DENTARY) of the lower jaw in vertebrates. Teeth situated on the dentary bone are termed 'mandibular teeth'.

MANDIBULAR Of, on, resembling, or close to the lower jaw or MANDIBLE of vertebrates.

MANDIBULAR TEETH *See* MANDIBLE.

MANGROVE SWAMP A tidal region typical of many tropical and, occasionally, subtropical sheltered coasts, frequently with freshwater lagoons and river deltas, and characterized by dense thickets of evergreen trees and shrubs of the genus *Rhizophora,* with stilt-like intertwining aerial roots. Mangrove swamps are important habitats for many amphibians and reptiles, including terrestrial, arboreal and sea snakes, certain varanid lizards and several *Crocodylus* species.

MARGINAL Of, on, in, or forming, a margin or border. **1**. Any one of the laminae appearing as a border around the edge of the bony part of the CARAPACE in chelonians. **2**. Any one of the scales in the upper of two rows lying laterally directly beneath and adjacent to the DORSAL scales in crocodilians.

MARINE Of, relating to, or inhabiting, the sea.

MARSH An area of low, waterlogged land, usually near or on the edge of lakes or rivers etc., and an important habitat for many species of reptile and amphibian, e.g., water snakes (*Natrix, Nerodia* etc.) and anurans and urodeles (*Rana, Triturus* etc.).

MARSUPIAL POUCH A sac-like cavity (marsupium) formed from the skin on the back of female frogs (genus *Gastrotheca*). The pouch opens directly above the CLOACA and receives the fertilized eggs which remain there until they have developed into free-swimming tadpoles (*Gastrothec.marsupiata*) or completely metamorphosed young frogs (*Gastrotheca. ovifera*).

MARSUPIUM *See* MARSUPIAL POUCH.

MASK 1. The dark, broad band or stripe, occurring in some auran species, running laterally from the nostril or the eye to the tympanum and, sometimes, beyond. **2**. The face-like markings, or pattern, on the back of the spread HOOD in several species of cobra.

MASSASAUGA; BLACK SNAPPER *Sistrurus catenatus*, a rattlesnake of the family VIPERIDAE, subfamily Crotalinae, inhabiting a variety of habitats from marshland to rocky desert hillsides, from the Great Lakes of Canada to northern Mexico. The name 'massasauga' comes from the Native American Chippewa tribe and is almost certainly derived from their name for the Missisauga River in Ontario, a likely environment for the rattlesnake.

MASTIGURE A large, heavy-bodied lizard of the genus *Uromastyx* in the family AGAMIDAE , inhabiting rocky desert areas of North Africa and the Middle East.

MATA MATA *Chelus fimbriatus,* a totally aquatic freshwater turtle of the family CHELIDAE, inhabiting water systems of northern South America.

MATRIX The fleshy tail tip in rattlesnakes (*Crotalus,* and *Sistrurus*), from which each successive horny ring, or segment, of the RATTLE develops.

MAXILLA (pl.) **MAXILLAE** The upper jaw bone of vertebrates, and in reptiles and amphibians either one of two, usually tooth-bearing, bones separated at the front of the jaw by the PREMAXILLA BONES. In viperid snakes the maxilla bears the fangs and in chelonians the teeth are

122

replaced by a horny covering
(BEAK).

MECKEL'S CARTILAGE In
reptiles and amphibians, the ossified
cartilage that forms the articular bone
of the lower jaw.

MEDIAL Of, or occurring in, the
middle.

MEDIAN Of, or occurring towards,
the middle.

MEDIAN EYE The PARIETAL
EYE of early vertebrates.

MEDIAN GULAR A scale situated
on the throat either between or
directly behind the CHINSHIELD
scales.

MEDIOGYRINID A term given to
anuran larvae which possess a single
SPIRACLE in the centre of their
bodies, e.g., the
LEIOPELMATIDAE and
DISCOGLOSSIDAE.

MEGADONT Having teeth that are
not regular in either size or
appearance.

MELANIN Pigment granules of the
skin responsible for dark brown to
black coloration in animals.

MELANISM An overproduction of
the pigment melanin resulting in
animals that are entirely, or nearly,
all-black, often as a response to
environmental factors. Melanism is
not uncommon among certain
species of snake and lizard, such
dark, or melanistic, individuals
sometimes receiving the term
'melanos'. The term is sometimes
wrongly used for ANERYTHRISTIC
individuals.

MELANISTIC *See* MELANISM.

MELANO *See* MELANISM.

MELANOPHORE A cell of the skin
containing the black or dark brown
pigment MELANIN.

MELANOSARCOMA *See*
TUMOUR.

MEMBRANE A very thin, sheet-like
tissue that connects, lines or covers
organs and cells of animals (and
plants). Examples are the
NICTITATING MEMBRANE that
covers the eye in many reptiles, and
the WEB that connects the digits in
many amphibians.

MENTAL The single MEDIAN scale
on the front edge of the lower jaw,
directly in front of the
CHINSHIELD in lizards, and
between the first lower LABIAL
scales in snakes.

MENTAL GLAND A gland situated
in the skin of the chin of certain male
urodeles, the secretion from which
stimulates the females into becoming
sexually responsive to mating.

MENTAL GROOVE The medial
furrow on the under surface of the
lower jaw between the
CHINSHIELD scales in snakes. It
allows the jaw, and the skin upon it,
to expand during the swallowing of
large food items.

MERCY CRY An alternative term
for the FRIGHT CRY of anurans.

MERCY-KILLING The humane
killing, or EUTHANASIA, of an
animal to relieve suffering from
disease or injury.

MERISTIC Of, or relating to, the
number of parts, organs or other
countable structures of an organism,
e.g., the scales of snakes.

MERREM, **Blasius**, (1761-1824), German zoologist and the first accurately to separate the reptiles and amphibians, in *Versuch eines Systems der Amphibien* (1820). He combined the snakes and lizards in a single order, the SQUAMATA, and also separated the crocodilians from the lizards.

MERTENS, **Robert, (**1894-1975), German zoologist and herpetologist, and Director of the Senckenberg Museum in Frankfurt am Main, Germany. His contribution to herpetology was tremendous with over 600 publications to his credit, a number of which he co-authored, including the definitive identification keys and checklists for the recent crocodilians, chelonians and tuataras, which he compiled with Wermuth (1955 & 1961).

MERTENSIAN MIMICRY The theory proposed by R. Mertens to explain the patterns of coral snakes (alternating black, yellow and red or white rings) that occur in both diverse and unrelated snakes. Such ringed coloration is thought to be a warning coloration and is believed to have originated among non-dangerous but rather aggressive species (e.g., genera *Erythrolamprus, Rhinobothryum*). They were able to communicate negative experiences to their enemies and therefore, by means of their coloration, they had a deterrent effect.

MESIC Of, relating to, or inhabiting, areas that are damp but not swampy.

MESO- A prefix meaning 'middle'.

MESOGLYPH Any one of a number of snakes in which the FANG is situated close to the middle of the relatively short maxillary bone of the upper jaw. To the front of the fang lie from two to ten smaller teeth, whilst to the rear of it there are no teeth at all. Examples of mesoglyph snakes are seen in the similar, and closely related, African genera *Polemon* and *Miodon*

MESOPLASTRON Either one of the pair of non-adjacent dermal bones situated between the HYOPLASTRON and HYPOPLASTRON bones of the PLASTRON, typical of certain primitive chelonian species but still surviving in the musk turtles (*Sternotherus*) of eastern North America, and the helmeted terrapins (PELOMEDUSIDAE) of Central Africa and Madagascar.

MESOPTYCHIAL Any one of the scales situated on the outer surface of the GULAR FOLD or on the rear, central area of the under surface of the throat (mesoptychis) directly in front of the PECTORAL GIRDLE in lizards.

MESOZOIC The middle geological era dating from the end of the PALAEOZOIC, approximately 230 million years ago, to the beginning of the CENOZOIC, approximately 70 million years ago. The dinosaurs, ichthyosaurs, pterosaurs etc. were the dominant animals at that time and the Palaeozoic is often termed the 'Age of Reptiles'. Most of them, however, had become extinct by the end of the era. The mesozoic era is made up of three main periods: the

124

TRIASSIC, JURASSIC and
CRETACEOUS.

METABOLIC BONE DISORDER; DECALCIFICATION; RICKETS A serious, progressive, debilitating and, if untreated, often lethal disease, usually resulting from insufficient dietary calcium and/or exposure to ultra violet light.

METABOLISM Chemical balance and reactions that occur in cells within the body, resulting in growth, energy production, removal of waste products etc. The rate of metabolism is the speed, higher in a warm-blooded animal (HOMOIOTHERM) than in a cold-blooded animal (POIKILOTHERM), at which the chemical reactions occur.

METACARPAL BONE Any one of the rod-shaped bones that form the METACARPUS in the forefoot or lower forelimb of tetrapod vertebrates, and articulate proximally with the CARPALS, and distally with the PHALANGES. In a typical pentadactyl (having five digits) limb there are five metacarpals although there are adaptations to this plan and, in many species, the number is reduced.

METACARPUS The collection of METACARPAL BONES that forms part of the skeleton of the forefoot or lower forelimb in tetrapod vertebrates.

METACHROMATISM A change in colour as a result in a change in the ambient temperature. Many reptiles snake and lizard species darken considerably when cooled.

METAMORPHOSIS The relatively rapid transformation occurring in an amphibian when the LARVA takes on the physical appearance of the ADULT. In anurans the most obvious sign is the loss of the tail and rapid growth of the legs. In urodeles it is the disappearance of the CAUDAL FIN and feathery gills.

METATARSAL BONE Any one of the rod-shaped bones that together form the METATARSUS in the hind foot or lower hind limb of tetrapod vertebrates, and articulate proximally with the TARSAL bones and distally with the PHALANGES. In a typical pentadactyl (having five digits) limb there are five metatarsals, although there are adaptations to this plan and, in many species, the number is reduced.

METATARSAL TUBERCLE A prominent knob-like projection present on the hind foot of many anurans and used by certain species (e.g., *Scaphiopus, Pelobates*) as a SPADE for digging.

METATARSUS The collection of METATARSAL BONES that forms part of the skeleton of the hind foot or lower hind limb in tetrapod vertebrates.

mg/kg The abbreviated expression for the quantity of venom required to kill an animal correlated to its mass, with 'mg' representing milligrams and 'kg' kilograms. The taipan *Oxyuranus scutellatus*, for example, has a venom toxicity of 0.064 mg/kg meaning that 0.064 mg (0.0064 grams) will be sufficient to kill an animal weighing 1 kg.

MIAMI PHASE A term applied to the eastern subspecies of the corn snake *Pantherophis guttatus guttatus* (formerly *Elaphe guttata guttata*) which is greyish with reddish-orange blotches.

MICRO- A prefix meaning small; minute.

MICROHABITAT Any habitat which is very small, such as that created within a hollow tree or beneath a wood pile, for example.

MICROHYLIDAE Narrow-mouthed toads. Family of the ANURA, inhabiting New Guinea, Asia, Africa, Madagascar, South America and North America. 279 species in 59 genera.

MID-BODY SCALE ROW 1. In snakes, the number of scales lying on an imaginary line running around the centre of the body, beginning and ending at, but not including, the large elongated VENTRAL scales. 2. In lizards, the number of scales lying on an imaginary line running completely around the centre of the body.

MID-DORSAL Of, relating to, or on, the centre of the back, or DORSUM.

MID-RIB The gristly mid-section of the vocal cord in certain anuran species.

MID-VENTRAL Of, relating to, or on, the centre of the abdomen, or VENTER.

MIGRATION *See* SEASONAL MOVEMENT.

MIGRATION ROUTE A connection between two biogeographical zones that allows the exchange of animals (and plants).

Several different categories are acknowledged: for example, G. G. Simpson's CORRIDOR DISPERSAL ROUTE, FILTER DISPERSAL ROUTE and SWEEPSTAKE ROUTE are extensively referred to in relation with mammalian, and more recently reptilian, migrations. *See* SEASONAL MOVEMENT.

MILK To extract the venom from venomous reptiles by skilful and manual means. Milking a venomous reptile is generally done by mechanically or electrically stimulating the venom glands to discharge their contents into a vessel over which the reptile has been persuaded to open its mouth.

MILKED-UP The term commonly used to describe a snake that is preparing to SLOUGH its skin which, together with the eye, assumes a pale bluish tinge caused by an oily secretion, from the EXUVIAL GLAND, that forms between the old and new skin, moistening the old one prior to ECDYSIS.

MILK SNAKE Originally, the harmless North and Central American colubrid, *Lampropeltis triangulum*, but used today for any of the tricoloured *Lampropeltis triangulum* subspecies.

MILT The spermatozoa and seminal fluid of a male anuran expelled, during AMPLEXUS, as the female simultaneously deposits her eggs which are then fertilized as the milt flows over them.

MIMIC Referring to any species that

closely resembles, or assumes the appearance of, a different, usually distasteful or venomous species. Among reptiles examples can be seen in the harmless tricoloured MILK SNAKES (*Lampropeltis triangulum* ssp.), which mimic the venomous coral snakes (*Micrurus* and *Micruroides*).

MIOCENE The fourth epoch of the TERTIARY period, beginning approximately 25 million years ago at the end of the OLIGOCENE, and lasting some 18 million years, when it was superseded by the PLIOCENE.

MITE A small arthropod of the order Acarina (which also includes the ticks). Many are parasites of plants and animals and some are vectors of disease. Mites exceed all other spider-like animals in their numbers and more than 20,000 different species, almost certainly just a small part of the total, are to be encountered in almost every type of habitat throughout the world. Most adult mite species have four pair of legs and a body which lacks any separation between head and abdomen. In parasitic mites, the mouthparts are usually specialized as piercing stylets. Larval trombiculid mites infect the skin of anurans, and the snake mite, *Ophionyssus natricus*, is particularly significant in reptile collections as it can act as a VECTOR of disease. Another mite occasionally encountered by reptile keepers who rear mealworms for their captives is the 'bran mite'. This is a yellowish sarcoptid mite of the

suborder Sarcoptiformes, which sometimes can infest stored products including the bran in which mealworms are cultured if it is allowed to become damp.

MOCCASIN A term frequently used incorrectly in referring to any water snake both venomous and non-venomous. The true moccasins however are dangerously venomous, North and Central American PIT VIPERS of the genus *Agkistrodon*. Known as 'water moccasins', or COTTONMOUTHS, they are represented in eastern North America by *Agkistrodon p. piscivorus,* and in western North America by *Agkistrodon p. leucostoma.* Their close relatives *Agkistrodo c. contortrix* and *Agkistrodon.. c. mokeson,* also of North America, are known as 'highland moccasins' or COPPERHEADS.

MOCQUARD, François, (1834-1971), French herpetologist. Contributor to the 17-volume *Mission scientifique au Mexique et dans l'Amerique Centrale: Etudes sur les reptiles et les batraciens* (1870-1909), in joint authorship with Brocchi, Duméril (Auguste) and Bocourt.

MOLCHPEST The feared and little understood disease that occurs, often in epidemic proportions, among captive urodeles, and is characterized by a variety of symptoms including dermal inflammation and the subsequent development of abscesses which spread rapidly over the body but, most significantly, by an unusual odour which the diseased amphibian, and the water in which it is housed, emits. Affected specimens invariably fail to respond to treatment and eventually die.

MOLOCH The thorny devil *Moloch horridus*, an Australian lizard of the family AGAMIDAE.

MOLT An alternative spelling for MOULT.

MONITOR Any one of the large, predatory lizards of the family VARANIDAE, inhabiting Africa, southern Asia and Australasia.

MONOCELLATE Possessing a single eye-like spot, or OCELLUS.

MONODACTYL Possessing a single digit on the fore- or hind foot, as in certain lizard species (e.g., SCINCIDAE).

MONOSPECIFIC **1.** Pertaining to a genus that contains a single species only. **2.** An alternative term for MONOVALENT.

MONOTYPIC Containing a single form (TAXON) only, e.g., a species in which there is no subspecies. *Contrast with* POLYTYPIC.

MONOVALENT Possessing antibodies for a single particular organism or substance, and a term used commonly for any antivenom that acts against the venom of a single snake species only.

MONSOONAL Descriptive of a climate pattern, prevailing over South east Asia, which has a wind system that changes direction with the seasons i.e., southwest in summer bringing heavy rains and northeast in winter.

MONSOON FOREST A type of deciduous forest of eastern and south-eastern Asia, that has a more or less well-determined monsoon climate.

MONTANE Of, relating to, or inhabiting, mountains or mountainous regions.

MONTANE DRY FOREST A type of forest, characterized by low deciduous trees, thick, thorny shrubs, and numerous types of grasses, occurring mainly between altitudes of 500 and 1500 metres.

MONTANE MOIST FOREST A type of forest, occurring mainly at altitudes above 1000 metres, in regions in which the seasons are well-determined, and receiving rainfall approximately intermediate in amount to that received by montane-dry and -wet forests.

MONTANE WET FOREST A type of extremely damp forest, occurring mainly at altitudes over 1500 metres and analogous in character to the highest parts of the CLOUD FOREST.

MORBIDITY The comparative occurrence of a specific disease in a particular area or locality; a tendency towards illness.

MORPH A form or phase; different from the normal, e.g., the European viper *Vipera berus*, in which the characteristic zigzag running from the back of the head to the tip of the tail is replaced by a straight-edged longitudinal stripe.

MORPHOLOGY The study of the phases, or forms, of organisms.

MORTALITY Death rate; the number of deaths within a given period.

MOTLEY A recently devised term for a corn snake *Pantherophis guttatus* with a genetic mutation resulting in abnormalities in the pattern of the dorsal blotches.

MOULT The act of sloughing, or casting off, the old transparent outer layer of skin during ECDYSIS.

MOUTH BROODING *See* BUCCAL INCUBATION.

MOUTH ROT *See* STOMATITIS.

MUCORMYCOSIS A highly infectious disease, prevalent in Australia, and caused by the fungus *Mucor amphibiorum* which affects the skin and internal organs of anurans. Whilst amphibian mucormycosis seems to occur only intermittently in wild populations of frogs and toads it can spread rapidly through captive collections resulting in high mortality.

MUCRONATE Ending in a sharp point, a term used to describe scales that are overlapping and taper into a spine.

MUCUS The slimy secretion produced by certain cells in the mucous membranes and glands of animals, the function of which is to lubricate and protect the surface upon which it is secreted. In many amphibians the mucous on the skin may be toxic to a varying degree, and this toxicity is particularly well-developed in that of the arrow-poison frogs of the family DENDROBATIDAE which, in some cases, can cause heart failure if it enters the blood stream.

MUDDLE To 'feel', with either the hands or feet, for aquatic chelonians as they lie hidden in the mud of dark or cloudy waters.

MUDPUPPY *Necturus maculosus*, a large aquatic salamander of the family PROTEIDAE, inhabiting streams, rivers and lakes in central and northern North America. The mudpuppies, together with the WATERDOGS, are permanently aquatic larvae characterized by deep red plume-like gills and strongly compressed tails.

MUGGER *Crocodylus palustris*, a large, freshwater crocodile inhabiting rivers, lakes and man-made waterways in the Indian subcontinent from eastern Iran, Pakistan, northern India and Nepal to Bangladesh and south to Sri Lanka. Also called the marsh crocodile.

MULGA *Pseudechis australis*, a large and formidable snake of the ELAPIDAE family, also known as the king brown snake, found in a variety of habitats in Australia, with the exception of the extreme south and New Guinea.

MÜLLER, Fritz, (1831-1897), German zoologist who was the first to recognize the type of mimicry named after him (MÜLLERIAN MIMICRY).

MÜLLER, Lorenz, (1868-1953), German herpetologist and contributor of numerous works on reptile and amphibian systematics, including the checklist *Die Amphibien und Reptilien Europas* (1928 and 1940) which he co-authored with Robert Mertens

MÜLLERIAN DUCT The female OVIDUCT of vertebrates which,

although it develops in both sexes and is linked to the WOLFFIAN DUCT, becomes vestigial in the male.

MÜLLERIAN MIMICRY The similarity in appearance of one species of animal to that of another, where both are distasteful to predators. Obvious examples are wasps and bees but the same principle is employed by several species of venomous coral snakes. All gain from having the same warning coloration, since predators learn to avoid both species after tasting either one or the other. The phenomenon is named after Fritz MÜLLER who described it in relation to insects in South America. *Compare* BATESIAN MIMICRY.

MULTICARINATE Possessing more than one KEEL, as in the CARAPACE of many chelonians.

MULTIVALENT An alternative term for POLYVALENT.

MUSCARINE-LIKE SYNDROME Low blood pressure (HYPOTENSION), slow pulse rate, dilated blood vessels, profuse sweating and increased PERISTALSIS, and a symptom in the case of certain venomous snake bites.

MUSCARINIC TOXIN A component of the venom of African mambas (*Dendroaspis* sp.) and which causes low blood pressure, dilated blood vessels and increased PERISTALSIS.

MUSK GLAND A gland secreting a strong-smelling, heady odour, used by male animals to stimulate the females into sexual activity. In crocodilians there are two pairs of musk glands, one pair situated at the corner of the jaw (the ANGULAR GLAND) and the other pair situated in the cloaca (the CLOACAL GLAND). Musk glands are also found in snakes (the ANAL GLAND) and in some chelonians.

MUSSURANA *Clelia clelia*, a rear-fanged snake of the family COLUBRIDAE, subfamily Boiginae, inhabiting Central and South America.

MUTANT; MUTATION 1. The act or process of mutating; alteration; change. **2.** An alteration in the chromosomes or genes of a cell. When this change occurs in the gametes the structure and development of the resultant offspring may be affected. **3.** An animal affected by the process of alteration.

MUZZLE Very occasionally used by some authors for the SNOUT of certain reptiles, with particular reference to crocodilians.

MYCOBACTERIAL DISEASE An infection caused by any rod-shaped Gram-positive bacterium of the *Mycobacterium* genus and in amphibians mainly affecting the respiratory system, intestines and skin. Mortality rates are generally low in otherwise healthy individuals, the infection chiefly occurring in already ill or weakened captive animals. In such individuals death may ensue once the infection reaches the kidney, liver and other internal organs.

MYCOSIS Any one of a number of infections or diseases caused by parasitic fungi, and affecting the mucous membranes of the internal organs, and skin, of reptiles and amphibians.

MYOBATRACHIDAE Family of the ANURA, inhabiting Australia. 106 species in 20 genera.

MYOCARDIAL Pertaining to the heart muscles.

MYOCARDITIS Inflammation of cardiac muscle tissue.

MYOFIBRIL Any one of the small fibres occurring in, and constituting muscle tissue.

MYOGLOBINURIA In reference to the symptoms of certain venomous SNAKE BITE, the passage of muscle pigment (myoglobin) in the urine.

MYOLTIC Causing destruction of muscle.

MYONECROTIC Causing the damage or death of muscle tissue

MYOTOXIC *See* MYOTOXIN.

MYOTOXIN A poison which attacks and damages the muscles, particularly the pulmonary (lungs) and cardiac (heart) muscles. Seasnakes possess myotoxic venoms.

MYRMECOPHAGUS Ant-eating.

N

NAIL The enlarged scale, or scales, forming a sharp, horny spine on the tail tip of certain chelonians, e.g., the European Hermann's tortoise *Testudo hermanni*.

NAKED Used in reference to the HEMIPENIS in some snakes, which lacks any surface decoration in the way of hooks or spines.

NAPE The back of the neck and, in snakes, referring to the dorsal area directly behind the head.

NARIS (pl.) **NARES** Either of the paired nasal openings, or nostrils, between the NASAL CAVITY and the exterior, consisting of the internal nares (or choanae), which open into the BUCCAL cavity, and the external nares, which open to the outside.

NASAL In reptiles, a scale, situated on the side of the head, that borders or contains a nostril (or naris). *See* POSTNASAL and PRENASAL.

NASAL CAVITY A paired cavity in the head of all vertebrates, lined with a mucous membrane rich in sensitive OLFACTORY receptors and connected by internal NARES to the respiratory system, and by external nares to the outside.

NASAL CLEFT The NASAL GROOVE in blind snakes of the TYPHLOPIDAE.

NASAL GLAND A gland occurring in the nostrils of certain terrestrial iguanids (e.g., *Ctenosaura, Sauromalus* and *Dipsosaurus*, and the marine iguana *Amblyrhynchus cristatus*) that discharges through a duct in the nasal cavity and serves as a means of excreting excess salt from the body. In crocodilians, marine turtles, and many other lizards, excess salt is excreted in the form of 'tears' from the LACRIMAL GLAND.

NASAL GROOVE In snakes of the family TYPHLOPIDAE, a cleft, or furrow, which in some species may be only partially developed, running from the PREFRONTAL scale, through the nostril, and over the NASAL scale to the lip.

NASAL VALVE A sphincter mechanism for shutting off the nostril and found in most sea snakes and some vipers.

NASOLABIAL In the thread snakes (LEPTOTYPHLOPIDAE) and blind snakes (TYPHLOPIDAE), the scale on the head containing the nostril, or NARIS.

NASOLABIAL GLAND Any one of a group of glands situated within, and serving to irrigate, the NASOLABIAL GROOVE of lungless salamanders (PLETHODONTIDAE).

NASOROSTRAL The small scale, or scales, situated on the head in certain pit vipers of the *Bothrops* genus, in the space between the ROSTRAL and the NASAL scales.

NATTERJACK *Bufo calamita,* the European running toad.

NATIVE *See* INDIGENOUS.

NATURA 2000 A project by the European Union and each of its Member States to protect the environment. Part of the project is the designation of special areas of biological importance such as, for example, SPECIAL AREAS OF CONSERVATION (SACs).

NATURAL SELECTION The process, which Darwin called the 'struggle for survival', by which those individual organisms best fitted for their environment survive, whilst those less well fitted do not, thus helping to ensure the survival of the species as a whole, individuals having unfavourable characteristics being 'selected against'. Natural selection, according to Darwinism, results in evolution when acting on a diverse population because it can lead to speciation by favouring a mutant if it displays features particularly advantageous to its mode of life.

NEARCTIC One of the six zoogeographical regions of the world, part of the HOLARCTIC, and consisting of North America from the Central Mexican Plateau in the south to the Aleutian Islands and Greenland in the north.

NECK BITE The bite to the NAPE of the neck in many species of snake and lizard and applied by the male to the female during copulation. The neck bite rarely results in any serious injury although, if copulation occurs frequently over a period of days or weeks the scales in the neck region might become quite badly marked.

NECROPSY An autopsy, or post-mortem examination, performed to ascertain the cause of death.

NECROSIS The death of one or more cells of the body, usually localized, and usually a result of a restricted blood supply to that part.

NECROTIC Causing death, as in tissue death occurring, for example, in snakes with necrotic DERMATITIS.

NECROTIC STOMATITIS *See* STOMATITIS.

NEKTON Pertaining to free-swimming animals that inhabit the middle depths of open water such as a lake or sea. The TADPOLE, or larva, of certain amphibians, that spends its period of development in such bodies of freshwater, is termed a 'nektonic tadpole'.

NEKTONIC Of, in or relating to the middle depths of a body of water.

NEKTONIC TADPOLE *See* NEKTON.

NEMATODA The class of both free-living and parasitic animals containing the roundworms and eelworms. Nematode worms occur as ENDOPARASITES in several internal organs, particularly the lung and alimentary canal, of both reptiles and amphibians.

NEMATODE Any unsegmented worm of the phylum NEMATODA, possessing a resilient outer cuticle. The group includes free-living forms and disease-causing parasites such as hook worms and filaria.

NEOLITHIC The recent Stone Age, originating in the Middle East approximately 10 000 years ago and lasting until the beginning of the

Bronze Age some 7000 years ago.

NEONATE A newborn, or hatchling reptile.

NEOTENE *See* NEOTENIC.

NEOTENIC Describing an organism or species that is sexually mature and capable of reproduction but retaining the appearance, form and habits of the larva. Such an animal is called a 'neotene'

NEOTENY The condition in which the larval features are retained in the adult due to the larva's failure to metamorphose. The axolotl *Ambystoma mexicanum* is probably the most well-known example of an animal species that engages in neoteny.

NEOTROPICAL Of, relating to, or inhabiting, any one or more of the tropical regions of the New World, consisting of South America and North America south of the tropic of Cancer. The Neotropical is one of the six main zoogeographical regions of the world.

NEPHRITIC Of, or relating to the kidneys.

NEPHRITIS Inflammation of the kidneys.

NEPHROSIS Any non-inflammatory degenerative kidney disease.

NEPHROTOXIC Causing damage to the kidneys.

NERITIC Of or formed in the region of shallow seas near a coastline between low-tide level to a depth of 200 m. *Compare* ABYSSAL; BATHYAL; and NERITIC.

NERVOUS SYSTEM A system of specialized cells and tissues in multicellular animals, in which information is communicated between receptors and effectors, and which allows the regulation and co-ordination of many body functions. In reptiles and amphibians (and in all other vertebrates) it consists of the central nervous system (the brain and spinal cord) and the peripheral nervous system (the cranial and spinal nerves and their branches).

NEST 1. Generally referring to the large mounds of soil, sand and vegetation created by crocodilians in which to lay and incubate their eggs. *See* FOAM NEST. **2.** A collective noun for toads, vipers, or snakes.

NEURAL; DORSAL; VERTEBRAL Any one of the central row of bones of the CARAPACE, in chelonians, corresponding to the central VERTEBRAL laminae. Anterior to the neurals is the PRONEURAL, posterior is the PYGAL.

NEUROLOGICAL Pertaining to the nervous system.

NEUROLOGY The study of the anatomy, physiology, and diseases of the nervous system.

NEUROLYTIC *See* NEUROTOXIC.

NEUROMAST One of many groups of cutaneous sensory cells, situated within pits or canals, either scattered or arranged in orderly rows on the body and head of aquatic amphibians (and most fishes). The sensory cells bear minute hair-like projections which are sensitive to changes in pressure, and can detect vibrations in the surrounding water too low to be perceived by the ear. *See* LATERAL LINE SYSTEM.

NEUROMUSCULAR TRANSMISSION The transmission of a stimulus from a nerve ending to its muscle.

NEUROPATHY Any disease of the nervous system.

NEUROPHYSIOLOGICAL Pertaining to physiology in relation to nerves.

NEUROTOXIC; NEUROLYTIC Describing the effects of certain snake venom components which are poisonous or destructive to the NERVOUS SYSTEM. *Compare* HAEMOTOXIC.

NEUROTOXIN Any constituent of snake venom that results in a partial or total breakdown of the NERVOUS SYSTEM, acting mainly by paralysis of the motor nerve cells which transmit impulses from the brain or spinal cord to various effector organs, such as glands or muscles including those controlling breathing. Consequently death may be caused through asphyxia (an inability to breathe). The effects of certain neurotoxins are more easily reversed than others. *See* POST- and PRE-SYNAPTIC NEUROTOXIN.

NEUTRALIZE The capability of a substance (e.g., ANTISERUM) to reverse the effects of another substance (e.g., VENOM).

NEWT Any one of the various small semi-aquatic urodeles, such as the palmate newt *Triturus helveticus* of Europe, which together with the salamanders and related forms constitute the order CAUDATA.

NEWT PLAGUE The epidemic-like, fatal disease of urodeles, also known as MOLCHPEST, and occurring frequently in captive animals, apparently through incorrect or inadequate husbandry.

NEW WORLD The Americas.

NICHE The status or functional role filled by an organism in a particular community, often called its 'ecological nich'. A particular organism's niche is determined by the food it eats, temperature tolerances, enemies etc. No two species can coexist in the same niche for competition would then occur and one would replace the other by natural selection.

NICTITATING EYELID *See* NICTITATING MEMBRANE.

NICTITATING MEMBRANE The third eyelid, in certain amphibians and reptiles, consisting of a thin, transparent fold of skin situated in the inner corner of the eye, which can be drawn across the cornea, beneath the moveable upper and lower eyelids, to clean and protect the surface without hindering the animal's vision.

NOAH Acronym denoting the National On-line Animal Histories (European version of SPARKS). Under the auspices of the National Federation of British Zoos, NOAH is a central database of records on all the wild life in United Kingdom zoological collections and is being used to plan the movement of animals between zoos in order to avoid the problems of in-breeding.

NOBLE, Gladwyn Kingsley, (1894-1940), American herpetologist and

Curator of Herpetology at the American Museum of Natural History in New York. *The biology of the Amphibia*, which he published in 1931, went on to become one of the definitive works on the natural history and biology of amphibians.

NOCTURNAL Active or occurring after dark.

NOD The repeated raising and lowering of the head in quick succession, observed in the males of many lizards, particularly iguanids such as *Anolis, Leiocephalis, Sceloporus* etc., when in the immediate proximity of females of the same species, and apparently a way of acknowledging their sex.

NODOSE Having nodes or knot-like bumps.

NOMENCLATURE A system of naming organisms, or groups of organisms, with a SCIENTIFIC NAME. In scientific work only one system is used, giving two names to a species, e.g., *Eryx conicus* (rough-scaled sand boa), and three to a subspecies, e.g., *Eryx conicus brevis*.

NOMEN CONSERVANDUM In taxonomy, a name which, under the regulations of the ICZN code, cannot be accepted but which is made suitable by means of particular procedures, using either the original or altered spelling.

NOMINATE RACE The first defined of a species upon which subspecies, if any, are based. For example, in the case of the rainbow boa *Epicrates cenchria*, the nominate race, or subspecies, is the Amazonian one *Epicrates cenchria cenchria*,

while the one occurring in Bolivia is named *Epicrates cenchria gaigei*. The name of a subspecies, like that of a species, always begins with a small initial letter. One subspecies under each species has a name that is the same as that of the specific name, and this is known as the nominate race.

NON-POISONOUS Lacking the effects or qualities of a poison; incapable of killing or inflicting illness. *See* POISONOUS.

NON-VENOMOUS Lacking the effects or qualities of a VENOM. Although a non-venomous animal can still inflict an extremely painful, and indeed serious bite often resulting in secondary infection, it is incapable of killing directly by its bite or sting. *See* VENOMOUS.

NOOSING A method of capturing lizards and crocodilians in which a loop of thread or rope (depending upon the size of the animal) is tied, usually with a slipknot, to the end of a pole or cane, and lowered over the head and neck of a basking animal and pulled abruptly upwards resulting in its safe restraint.

NOSE KNOB The prominent hump, or GHARA, on the tip of the snout of a mature male gharial.

NOSTRIL Either one of the pair of external openings to the air passages, situated at the end of the snout.

NOTCH *See* AXILLARY NOTCH; INGUINAL NOTCH.

NOTCHED Having a V-shaped nick or indentation.

NOTCHED DIGITAL DISC Referring to the toe pad of certain arboreal frogs, which is not circular as in most other species but indented around its anterior edge.

NOTOCHORD Characteristic of CHORDATES, the notochord is a stiff cartilaginous rod of tissue extending the length of the body. In vertebrates, the backbone is deposited around the notochord and nerve cord.

NOTOGAEA The name given to the two combined southern zoogeographical regions of the world, the AUSTRALASIAN and the NEOTROPICAL.

NPS Acronym denoting the National Park Service (USDI).

NSHP Abbreviation for NUTRITIONAL SECONDARY HYPERPARATHYROIDISM.

NUCAL A variant spelling of NUCHAL.

NUCHA (pl.) **NUCHAE** Pertaining to the back, or NAPE, of the neck, e.g., a scale on the back of the neck of a crocodilian is termed a 'NUCHAL SCALE'.

NUCHAL BONE The PRONEURAL bone of a chelonian carapace.

NUCHAL CREST A longitudinal, central series of enlarged scales or fold of skin on the back of the neck in many lizards.

NUCHAL LAMINA The central front lamina (the precentral) overlying the PRONEURAL bone in a chelonian carapace.

of the usually two rows of enlarged, osteoderm-reinforced scales on the back of the neck. **2.** In certain lizards, any one of the scales situated

on the back of the neck directly behind the head.

NUCHAL POISON GLAND *See* NUCHAL VENOM GLAND.

NUCHAL SHIELD A single, small lamina, or scute, situated medially between the frontmost pair of MARGINAL laminae on the carapace of a chelonian.

NUCHAL VENOM GLAND; NUCHAL POISON GLAND

NUCHODORSAL GLAND Any one of a number of glands situated beneath the skin of the NAPE region in several Asiatic water snakes (e.g., various Natricinae, *Macropisthodon* and others), the secretion from which is of uncertain function and only released if the skin covering the gland is broken, when it can cause irritation if allowed to come into contact with the mucous membranes.

NUCHOMARGINAL Any one of the group of small scales arranged around the NUCHAL SCALES on the back of the neck in crocodilians.

NUPTIAL EXCRESCENCE *See* NUPTIAL PAD.

NUPTIAL BRUSH *See* GLOVE.

NUPTIAL PAD; NUPTIAL EXCRESCENCE In male anurans, an area of usually roughened and darkly pigmented skin developing prior to and during the breeding season, and usually, but not always, disappearing once breeding has ceased. The nuptial pad is usually evident on certain digits, particularly the thumb, but can also occur on areas of the chin, chest and under surfaces of the limbs etc., and serves to maintain a firm hold on a female during AMPLEXUS.

NUTRITIONAL SECONDARY HYPERPARA-THYROIDISM A condition arising in captive reptiles as a direct result of an excess of phosphorus in the diet causing the bones to weaken. It can also arise from an incorrect calcium: phosphorus balance, or insufficient calcium in the diet. In juvenile and subadult reptiles NSHP is more commonly known as rickets. When the condition occurs in adult animals it is known as OSTEOMALACIA.

NYIKA In eastern and southern Kenya east of the rift valley, a region of low, dry savanna extending from the coastal plain to the mid-altitude moist savanna.

NYMPH A term generally referring to the larval form of certain insects, such as the dragonfly, which develops into an adult without going through a pupal stage. In herpetology, it is used occasionally for the immature stage, or TADPOLE, of urodeles.

NYSTAGMUS In reference to the effects of certain venomous snake bites, the involuntary movement of the eyeball.

O

OAR The vertically compressed tail of a sea snake.

OASIS (pl.) **OASES** A lush area in a desert occurring where the water table comes near to, or reaches, the ground surface. Such fertile places in otherwise barren and inhospitable wastelands may be few and far between and, consequently, are havens for a variety of desert-dwelling animals including several species of reptiles and amphibians.

OBLIQUE TYPE In the scalation of snakes, a particular arrangement in which the rows of scales run at an angle of approximately 30° to the horizontal plane. The majority of species have scales in rows running approximately 45° to the horizontal plane. *See* SCALE ROW.

OCCIPITAL Of, or relating to, or situated on, the occiput, or back of the head or skull.

OCCIPITAL CONDYLE A single or paired, rounded, bony prominence that protrudes from the back (OCCIPUT) of the tetrapod skull and articulates with the first cervical vertebra (the ATLAS), allowing the raising and lowering of the head. In amphibians there are two and in reptiles just one.

OCCIPITAL LOBE In certain members of the CHAMAELEONIDAE a fleshy flap or lobe on the back of the head.

OCCIPITAL SCALE Occasionally used in reference to the scale containing the vestigial PARIETAL EYE, situated directly behind the PARIETAL scales (and INTERPARIETAL scales if present) in many lizards.

OCCIPUT The hindmost part of the head or skull of vertebrates, where it joins the vertebral column..

OCEANIA The islands of the central and South Pacific, including Polynesia, Melanesia, and Micronesia: sometimes also including the Malay Archipelago and Australasia..

OCEANIC Of, relating to, or inhabiting, the regions of the sea beyond the continental shelf, with depths greater than 200 m. The sea snakes of the Hydrophinae, a subfamily of the ELAPIDAE, are true oceanic reptiles, spending the whole of their lives in a marine environment.

OCELLATED Having eye-like spots, or OCELLI.

OCELLUS (pl.) **OCELLI** An outlined eye-like marking; any one of the circular, eye-like spots consisting of two or three concentric light and dark rings around a contrastingly coloured central spot, occurring in the skin patterns of some reptiles, particularly lizards, and certain amphibians.

OCULAR 1. Any one of the scales forming the border of the ORBIT of the eye. the POSTOCULAR, PREOCULAR, SUBOCULAR and SUPRAOCULAR scales refer respectively to those lying behind, in

front of, below and above the eye. In blind snakes (family TYPHLOP-IDAE), the ocular scale actually covers the eye which lies, barely visible as a dark spot, beneath it.
2. An enlarged and irregular scale situated between the SUPRACILIARY and SUPRAOCULAR scales on the head of a crocodilian.

OCULAR PARALYSIS *See* OPTHALMOPLEGIA A paralysis of the muscles responsible for eye movement and resulting in double vision.

OCULOLABIAL In blind snakes (TYPHLOPIDAE), the cephalic scale originating from the union of the OCULAR (sense 2) with the LABIAL directly beneath it.

ODONTOID Resembling a tooth. Odontoid structures occur on various bones in the mouth in certain anurans and, in the aggressive African bullfrog *Pyxicephalus adspersus*, they are large enough to inflict serious bites.

ODONTOID PEG *See* ODONTOID PROCESS.

ODONTOID PROCESS; ODONTOID PEG The tooth-like, upward projection on the second cervical vertebra (the axis) of the vertebral column, which in reptiles (and mammals and birds also) articulates with the first cervical vertebra (the atlas), allowing side-to-side movement of the head.

OEDEMA; EDEMA The excessive accumulation of serous fluid in the intercellular spaces of tissue.

OESOPHAGEAL TEETH The highly modified, enamel-coated HYPAPOPHYSIS of egg-eating snakes (Dasypeltinae), used to break the shells of eggs as they are swallowed.

OESOPHAGUS The muscular tube in the alimentary canal that links the BUCCAL cavity or pharynx with the stomach. The mucous membrane lining is folded, permitting great expansion during the passage of food which is directed down to the stomach by rhythmic, wavelike contractions of circular and longitudinal muscle layers situated along its length.

OGADEN A region of South Eastern Ethiopia consisting of a low altitude, dry savanna.

-OIDEA A standardized suffix used to indicate an animal SUPERFAMILY in the acknowledged code of CLASSIFICATION. The Colubroidea for example, includes the Viperidae FAMILY and the Viperinae SUBFAMILY.

OKADA, Yaichiro, (1892-1976), Japanese ichthyologist and herpetologist, regarded as the father of herpetology in Japan. He published many works on both the fishes and the reptiles and amphibians of that country, including the classic *The tailless batrachians of the Japanese Empire* (1931).

OKEETEE A term sometimes applied to the eastern subspecies of the corn snake *Pantherophis guttatus guttatus* (formerly *Elaphe guttata guttata*) which is predominantly orange in colour

OLD WORLD The continents of the world known before the discovery of the Americas, comprising Europe, Africa and Asia.

OLFACTION *See* OLFACTORY.

OLFACTORY **1**. Of, or relating to, the sense of smell (olfaction). **2**. One of a number of nerves or organs (e.g., JACOBSON'S ORGAN) connected with the sense of smell.

OLIGO- A prefix meaning little; few, lack of.

OLIGOCENE The third epoch of the TERTIARY period, beginning about 38 million years ago, after the EOCENE epoch, and lasting for about 13 million years when it was followed by the MIOCENE epoch. It was characterized by the gradual disappearance of earlier mammals and their replacement by more modern groups such as the first pigs, tapirs and rhinoceroses.

OLIGODONT Possessing only a few teeth, each one spaced well apart from the other.

OLIGOPHYDONT Possessing a number of sets of teeth which are shed periodically throughout an animal's life culminating with one full set of permanent teeth. Characteristic of many lizards and crocodilians.

OLM *Proteus anguinus*, the European cave-dwelling salamander of the family PROTEIDAE.

OMNIVORE An animal that feeds on both animal and vegetable matter. Omnivorous reptiles are to be found in many lizard and chelonian families. Although the larvae of many amphibians eat plant matter at some stage in their development, the adults are strictly carnivorous, except for one species of South American tree frog *Hyla truncata*, which besides insects includes the fruits of certain trees in its diet, thus having the distinction of being the only known truly omnivorous amphibian.

ONTOGENESIS; ONTOGENY The entire course of development of an individual organism, from the fertilized egg through to adulthood. *Compare* PHYLOGENESIS.

ONTOGENETIC CHANGE Any change in patterning, coloration or morphology with increasing age or maturity, i.e. between juvenile and adult. Emerald tree boas *Corallus canina* and green tree pythons *Morelia viridis* exhibit ontogeneric colour changes from juvenile to adult.

OOCYTE An egg prior to maturation or OOGENESIS. In the ovary of female animals, any one of a large number of reproductive cells that has the potential to become an OVUM. In snakes, for example, oocytes are being formed continually by germinal tissue cells within the OVARY. The ovarian FOLLICLE then matures and releases OVA which are in turn fertilized by sperm.

OOGENESIS The formation, growth and maturation of OVA (egg cells) in the ovary of female animals.

OOPHAGOUS Egg-eating.

OPEN POPULATION A population that is freely exposed to GENE flow. *Compare* CLOSED POPULATION.

OPERCLE Technical term given to the EAR FLAP of a crocodilian.

OPERCULAR FLAP *See* EAR FLAP.

OPERCULAR GILL In anuran larvae, the GILL enclosed beneath the OPERCULUM. Referred to occasionally, but incorrectly, as the INTERNAL GILL.

OPERCULUM A lid-like flap, or process, covering an aperture; a cover of skin in anuran larvae (tadpoles) that grows back over the gills, completely enclosing the gill chamber with just a single opening to the exterior, the SPIRACLE.

OPHIDIA The snakes. A suborder of the SQUAMATA. (Also known as the SERPENTES).

OPHIDIAN **1.** Having the appearance of a snake. **2.** Of, associated with, or belonging to the OPHIDIA. **3.** Any reptile of the suborder OPHIDIA.

OPHIOLOGY The branch of zoology dealing with the study of snakes.

OPHIOLOGIST A person who studies snakes.

OPHIOPHAGOUS *See* OPHIOPHAGY.

OPHIOPHAGY The eating of snakes. Any organism that habitually includes snakes in its diet is termed 'ophiophagous'. Many reptiles, and even large anurans, will feed upon snakes if given the opportunity but certain snakes themselves have a preference for snakes over other prey and will even devour their own kind. The king snakes (*Lampropeltis* sp.), king cobra *Ophiophagus hannah*, indigo snake *Drymarchon corais* and kraits (*Bungarus* sp.) are some of the more well-known ophiophagous snake species.

OPHIOTOXICOLOGY The branch of science concerned with the venoms of snakes, their nature, effects and antidotes.

OPISTHODONT *See* TEETH.

OPISTHOGLYPH Any venomous snake in which the venom-conducting teeth are situated towards the rear of the mouth on the upper jaw (MAXILLA). A rear- or back-fanged, snake. A number of workers now use the term ECTOGLYPH (sense 2).

OPISTHOGLYPHIC TEETH The venom-conducting fangs of back-fanged colubrid snakes, on which the grooves are situated either anteriorly or posteriorly.

OPLURIDAE Malagasy lizards. Family of the SQUAMATA, suborder SAURIA, formerly of the IGUANIDAE, inhabiting Madagascar and the Comoro Islands. Some seven species in two genera.

OPTHALMIA Inflammation of the structures within the eye.

OPTHALMOPLEGIA; OCULAR PARALYSIS In reference to the symptoms of certain venomous SNAKE BITES, a paralysis of the muscles which control eye movement.

OPTIMAL FORAGING Any feeding activity which uses the least effort for acquiring the most food in the least amount of time

OPTIMUM TEMPERATURE RANGE The particular limits within which a reptile or amphibian can function normally. In reference

142

to captive husbandry, the most satisfactory range of temperatures, between the VOLUNTARY MINIMUM and VOLUNTARY MAXIMUM, at which a given species should be housed (generally equal to those temperatures experienced by the animal in the wild) in order for it to continue its natural functions and habits. *See* CRITICAL MAXIMUM; CRITICAL MINIMUM; VOLUNTARY MAXIMUM; VOLUNTARY MINIMUM.

ORAL DISC The area around the mouthparts of anuran larvae (tadpoles), consisting of the mandibles (BEAK), lips (labia), and the arrangement of horny, rasping cusps. The arrangement of these structures varies with the species and is therefore taxonomically significant.

ORAL FESTOON *See* ORAL PAPILLAE.

ORAL PAPILLAE; LABIAL PAPILLAE In anuran larvae, the numerous small, nipple-like projections that commonly form a fringe (the oral festoon) encircling the mouth, functioning as tactile, and possibly also chemical, receptors.

ORAL SUCKER *See* ADHESIVE ORGAN.

ORBICULAR Spherical, rounded, or disc-shaped.

ORBIT The bony socket of the eye; the border of skin around the eye of a reptile or amphibian.

ORBITAL An alternative term for the OCULAR (sense 1) scale on the border of the eye in reptiles.

ORBITAL APPENDAGE Used in reference to the fleshy, hornlike structures above the margin of the upper eyelid in certain anuran species, e.g., *Ceratophrys cornuta* and *Megophrys nasuta*.

ORDER The taxonomic category used in the classification of organisms that consist of one or more similar families. Similar orders form a CLASS and may be separated into suborders. Zoological order names typically end in '-a', as for example, SQUAMATA (snakes, lizards and amphisbaenids) and ANURA (frogs and toads).

ORDOVICIAN The second oldest geological period of the PALAEOZOIC era, beginning about 510 million years ago at the end of the CAMBRIAN period, and lasting for about 70 million years when it was superseded by the SILURIAN period. It is characterized by being almost completely lacking in vertebrates, with the exception of certain jawless fishes, although marine invertebrates were particularly abundant.

ORGAN Any distinct part of an organism consisting of a group of different tissues specialized to perform particular vital functions. Examples in reptiles and amphibians include the INTROMITTENT ORGAN of copulation, the kidney, lung, eye, ear etc.

ORGANISM An individual living animal or plant capable of maintaining the processes characteristic of life, especially reproduction.

ORGAN OF LEYDIG The vestigial third eye (the PARIETAL EYE) of reptiles.

ORGANOGENESIS The formation and developmental progress of organs within an organism.

ORIENTAL REGION One of the six zoogeographical regions of the world, including the tropical southern Asian countries of India and Sri Lanka, to Borneo, Java and the Philippines. The boundary between the Oriental and Australasian regions has caused much argument in the past. *See* WALLACE'S LINE.

ORIENTATION The ability of an animal to determine its exact position with regard to the points of a compass, or its adjustment to its surroundings or other external stimuli. Orientation occurs, for example, as a reptile or amphibian alters the position of all, or part, of its body in response to heat or light.

ORNAMENTATION Used in reference to all the many varied embellishments or anatomical structures that occur on the limbs and bodies of many reptiles and amphibians, e.g., CREST; DEWLAP; FIN; HORN; SPINE etc.

OROPEL A local name, used in parts of Central America, for the yellow colour morph of the eye-lash viper, *Bothriechis schlegelii.*

ORTHO- A prefix meaning straight; correct or right.

ORTHOKINESIS The movements of an animal in reaction to a stimulus, such that the speed of movements is relative to the power of the stimulus.

OS INNOMINATUM In adult reptiles, either one of the two lateral halves of the single bone resulting from the fusion of the ISCHIUM, ILIUM, and PUBIS bones of the pelvis.

OSMOREGULATION The process by which animals control the water content and concentration of salts within their bodies. In reptiles, salt excess in the body is excreted via certain SALT GLANDS such as the LACRIMAL and NASAL GLAND.

OSSICLE Any small, often irregularly-shaped bone in the body of an animal, especially one of three in the middle ear.

OSSIFICATION The formation of bone, or the actual process of being converted into bone. The bony abdominal ridges in crocodilians are ossifications of the skin.

OSTEODERM A bonelike dermal plate or area beneath the epidermal scales in some reptiles, e.g., crocodilians, helodermatid lizards etc., giving added protection and strength to the skin.

OSTEOMALACIA A disease occurring in captive adult reptiles characterized by a softening of the bones and resulting from a deficiency of calcium and phosphorus or vitamin D, or a combination of all three. *See* NUTRITIONAL HYPERPARATHYROIDISM.

OTIC In, on, or near the ear.

OUTBREEDING Breeding between entirely unrelated or only distantly related individual animals of the same species. Crossing between one

144

species and another (HYBRIDIZATION) generally results in sterile offspring and there are, consequently, various behavioural mechanisms to discourage it. Outbreeding increases the number of heterozygous individuals, thereby producing populations with more variation and adaptability towards changes in the environment.

OVA *See* OVUM.

OVARIOSALPINGECTOMY The surgical removal of the OVARY and OVIDUCT.

OVARY (pl) **OVARIES** The female reproductive organ which produces eggs, or ova.

OVARIECTOMY The surgical removal of either or both ovaries

OVIDUCT; MÜLLERIAN DUCT The tube, or duct, that carries the eggs from the OVARY to the CLOACA.

OVIPARITY Reproduction involving the production of undeveloped eggs, within membranes or shells, that are spawned or laid by the female. The entire development of the embryos, which are nourished in their eggs by large yolk sacs, occurs outside of the mother's body. Reptiles and amphibians reproducing this way are termed 'oviparous'.

OVIPOSITION The process, or act, of depositing eggs.

OVIPOSITOR The egg-depositing structure situated at the posterior of the abdomen in female insects, and certain fishes, from which eggs are extruded often into, or beneath,

otherwise inaccessible places. In herpetology, the term is applied to the extension of the OVIDUCT in a few female anuran species, e.g., the Surinam toad *Pipa pipa*, which is protruded from the body and used to deposit individual eggs in each of the specialized dermal brood pockets on their backs.

OVOTHECA A widened section of the female anuran OVIDUCT in which the unfertilised eggs are held for a short time prior to AMPLEXUS.

OVOVIVIPARITY Reproduction involving the production of eggs which have a well-developed membranous covering and a store of yolk for the embryos' nourishment, are retained within the female's body during embryonic development, and hatch immediately before, during, or just after, they are laid. Reptiles and amphibians reproducing this way are termed 'ovoviviparous'. *Compare* OVIPARITY; VIVIPARITY.

OVULATION The release of an egg (OVUM) from the OVARY.

OVUM (pl) **OVA** A mature reproductive cell (GAMETE) in female animals, produced by the ovary and capable of developing into a new individual of the same species when fertilized by sperm from a male.

OXYURIDAE A family of endoparasitic nematode worms, known commonly as threadworms, containing some 150 species, a number of which infect the intestinal mucous membranes of reptiles, especially chelonians.

P

P *See* PARENTAL GENERATION.

PADDLE The name for the flattened limb of a marine turtle, specialized for swimming.

PADDLE STAGE The phase in the life of larval anurans in which the limb bud becomes a paddle-like form, culminating in the appearance of recognizable features of the foot.

PAEDOGENESIS Reproduction by an animal whilst still in the pre-adult, or larval, form. A type of NEOTENY, paedogenesis occurs, for example, in the axolotl *Ambystoma mexicanum*, a larval form of salamander which, although retaining larval features, such as gills and fins, can breed and produce offspring similar to itself.

PAEDOMORPHOSIS Evolutionary change that results in the withholding of juvenile characteristics into adult life. A state occurring in certain amphibians, particularly urodeles, in which the adult form retains larval characteristics e.g., external gills

PALAEARCTIC REGION One of the six zoogeographical regions of the world, comprising the whole of Europe, the former USSR, Africa north of the Sahara, and most of Asia north of the Himalayas.

PALAEOCENE The oldest geographical epoch of the TERTIARY period, beginning some 65 million years ago at the end of the CRETACEOUS period, and lasting for about 11 million years until the beginning of the EOCENE period, and characterized by the absence of many reptiles and the presence of numerous primitive mammals now extinct.

PALAEOHERPETOLOGY The study of EXTINCT reptiles and amphibians, including their anatomy, classification (TAXONOMY), history and behaviour and embracing the fields of ecology, geology, palaeontology, zoogeography, and computer science.

PALAEOZOIC The first and oldest era in which life forms became diverse and abundant, beginning about 580 million years ago at the end of the PRECAMBRIAN, and lasting for some 350 million years when it was succeeded by the MESOZOIC era. It consists of the Lower Palaeozoic (CAMBRIAN, ORDOVICIAN, and SILURIAN periods) and the Upper Palaeozoic (DEVONIAN, CARBONIFEROUS and PERMIAN periods).

PALATAL TEETH The innermost row of teeth situated on the PALATINE and PARASPHENOID dermal bones on either side of the PALATE.

PALATE The roof of the mouth. Skin covering the palate in amphibians is modified as a surface for respiration.

PALATINE Either one of the pair of, in some species, tooth-bearing bones in the roof of the mouth, situated directly behind the VOMER which forms part of the palate.

PALATINE TEETH A term used generally for any teeth situated on any of the bones of the palate, e.g., VOMERINE TEETH but in snakes referring to those teeth situated on the PALATINE bone.

PALLOR A pale condition of the skin and mucous membranes usually suggestive of low blood pressure or anaemia, both symptoms in certain cases of venomous snake bite.

PALMAR Of, or relating to, the palm or ventral surface (face) of the hand or forefoot.

PALMAR TUBERCLE A small, knob-like projection on the ventral surface, or palm, of the hand or forefoot in anurans.

PALMATE Digitate; shaped like an open hand, with parts diverging from a common base, each connected by a web. The European urodele (*Triturus helveticus*) is named the 'palmate newt' after the webbing between the digits of the hind feet of the male during the breeding season.

PALPATION A method used to locate either ovarian FOLLICLES or ova in a female snake. The snake is allowed to move slowly over the palms of the hands whilst it is grasped gently but firmly. Follicles or ova, if present, can be felt in the snake's VENTER as they pass across the palms.

PALPEBRA The eyelid.

PALPEBRAL Of, or relating to, the eyelids, as in 'palpebral membrane' (the transparent eyelid in many anurans), or any of the numerous small scales on the upper eyelids of lizards (the SUPRAOCULAR

scales), for example.

PALPEBRAL APPENDAGE The ORBITAL APPENDAGE of certain anurans, e.g., *Ceratophrys cornuta*.

PALPEBRAL DISC The transparent spectacle on the lower eyelid of certain skinks, e.g., *Leiolopisma*, permitting vision when the eyes are closed.

PALPEBRAL MEMBRANE The transparent eyelid in many anuran species, e.g., certain ranids and hylids, serving to protect the eye whilst still permitting vision.

PALSY Paralysis resulting from damage to the nervous system.

PALUDARIUM Used occasionally in reference to a vivarium in which a tropical rainforest habitat has been re-created, with densely planted epiphytes and bromeliads etc., and, in larger facilities, small trees and shrubs, and also containing an area of water for more aquatic reptiles and fishes.

PALUSTRAL; PALUSTRINE Of, related to, or inhabiting, marshes or swamps.

PANOPHTHALMITIS A slowly progressive disorder in snakes which are kept in suboptimal conditions, occurring either as an isolated condition or supplementing other bacterial diseases. An obstruction of the LACRIMAL DUCT system of the eye, it is caused by GRAM-NEGATIVE BACTERIA spreading up through the lacrimal duct into the area between the BRILLE, or transparent scale covering the eye, and the cornea, resulting in a characteristic inflammation, discoloration, and swelling of the eye, causing blindness and necessitating its surgical removal if the snake is to survive.

PAPILLA (pl.) **PAPILLAE** A thin, small, elongated and often flexible, fleshy projection on or in the body of an animal that may be tactile in function.

PAPILLOMATA A form of dermal tumour often seen in the skin of certain lizards. *See* POX.

PARACHUTE The free, membranous flaps and frills on the sides of the head, trunk, tail and limbs, and between the toes, of gekkonid lizards of the genus *Ptychozoon*, and the wide webbing between the toes in frogs of the genus *Rhacophorus*, permitting them to slow their descent when leaping or gliding from one tree branch to another or to the ground.

PARAESTHESIA In reference to the symptoms of certain venomous snake bites, an unusual sensation such as burning, pricking or numbness.

PARALYTIC ILEUS In reference to the symptoms of certain venomous snake bites, the paralysis of the small intestine.

PARAMYXOVIRUS A highly contagious viral disease affecting snakes and, in particular, members of the families ELAPIDAE and VIPERIDAE. The characteristic symptoms include gaping, lethargy, refusal to feed, and rapid degeneration of muscle tone.

PARAPATRIC Said of species which have adjoining but separate habitats. *Compare* ALLOPATRIC; SYMPATRIC.

PARAPINEAL EYE *See* PARIETAL EYE, PINEAL COMPLEX.

PARAPINEAL ORGAN *See* PINEAL COMPLEX.

PARAPSID Descriptive of a type of skull found in extinct Ichthyopterygia, characterized by a single, high temporal opening and considered to be a variant of the EURYAPSID skull.

PARASITE Any organism that derives its nourishment from another (the HOST). In reptiles and amphibians such an organism may occur on the skin (an ECTOPARASITE) or in the body (an ENDOPARASITE). Heavy infestations may lead to disease in, and eventual death of, the host.

PARASITOLOGY The study of parasitic organisms which dwell on or within other creatures (HOSTS).

PARASTHESIA In reference to the effects of certain venomous SNAKE BITES, an uncharacteristic sensation such as burning, numbness or pricking.

PARATOID A variant spelling of PAROTOID.

PARATYPE In taxonomy, a specimen other than a TYPE specimen that is drawn on, and nominated as such, by an author at the time of the initial description.

PARAVENTRAL Any one of the scales making up the series that run longitudinally on either side of the VENTRAL scales in snakes.

PARAVERTEBRAL Pertaining to one or both sides of the vertebral, or MID-DORSAL, in reference to scales or markings etc. A stripe, for example, lying on either side of and paralleling the vertebral line of the back is termed a 'paravertebral stripe'.

PARAVERTEBRAL STRIPE *See* PARAVERTEBRAL.

PARENTAL GENERATION; P The generation consisting of the immediate parents of the F_1 GENERATION. Second and third generations may sometimes be denoted by the symbols P_2 and P_3 respectively.

PARENTERAL The method by which drugs are administered other than by mouth. In the case of reptiles, drugs administered parentally are given subcutaneously (i.e., beneath the skin).

PARESIS A partial or slight paralysis, sometimes noted as a muscle weakness.

PARIETAL 1. Either one of a pair of large scales situated on the top of the head in snakes, directly behind the FRONTAL and lying over the parietal bones that form the main part of the roof of the skull. 2. Any one of the large scales situated on the top of the head in lizards, behind the FRONTOPARIETAL plates (sense 2). 3. The area of the head in snakes and lizards, occupied by many small scales which replace the large, paired parietal scales in certain species.

PARIETAL EYE; FRONTAL ORGAN; MEDIAN EYE; ORGAN OF LEYDIG;

PARAPINEAL EYE; PINEAL EYE; PINEAL ORGAN; THIRD EYE The vestigial sensory structure, or 'third eye', that develops from an outgrowth of the forebrain of early vertebrates, situated in the centre of the skull between the PARIETAL scales. *See* PINEAL COMPLEX.

PARKER, Hampton Wilder, (1891-1968), British herpetologist at the British Museum (Natural History), London. Publications include the monograph *Frogs of the family Microhylidae* (1934), and the standard handbook *Snakes: a natural history* (1977) which he co-authored with A.G.C. Grandison.

PAROTID GLAND A comparatively large, serous, salivary gland in many mammals and, in man, lying in front of and below each ear. In certain colubrid snakes the gland has become modified for the production of venom and is often associated with the enlarged grooved teeth of rear-fanged species, in which it is known as DUVERNOY'S GLAND.

PAROTOID GLAND Either one of the pair of large, external, wart-like, toxin-secreting glands situated behind the eye, on the neck, or around the shoulder region, in many anurans and certain urodeles (e.g., *Bufo* and *Salamandra*).

PAROXISM A sudden spasm, or convulsion.

PARTHENOGENESIS Reproduction, without fertilization by a male, in which an unfertilised ovum develops directly into a new individual; virgin birth.

Parthenogenesis occurs in several reptile genera, including the blind snake *Ramphotyphlops bramina* and has been observed in some forms of the urodele genus *Ambystoma*.

PARTHENOGENETIC Any organism that reproduces without a male element, i.e., that engages in PARTHENOGENESIS; existing as female only populations capable of producing fertile eggs without the presence of a male. Several lizards are known but only a single snake, the Brahminy blindsnake *Ramphotyphlops braminus,* is the only known parthenogenetic snake but several neotropical lizard genera, *Cnemidopherus, Gymnophthalmus* etc., contain parthenogenetic species.

PARTURITION The process or act of giving birth or laying eggs.

PATAGIAL RIB Any one of the five to seven pairs of false ribs in the flying lizards (genus *Draco*) used to support the fan-like wing, or PATAGIUM.

PATAGIUM (pl.) **PATAGIA** The membranous web of skin between the limbs and body that serves as a wing in gliding animals.

PATELLA In certain reptiles, a bone situated in the hind limb anterior to the knee joint and set in a ligament of the muscles which straighten the limb. The knee-cap.

PECK ORDER *See* DOMINANCE.

PECTINATE Shaped like the teeth of a comb, as in the feathery gills of the axolotl *Ambystoma mexicanum*.

PECTORAL 1. Of, or relating to, the area of the body where the forelimbs originate; the chest, breast, or thorax.

2. Either one of the third pair of laminae on the PLASTRON of a chelonian.

PECTORAL AMPLEXUS; AXILLARY AMPLEXUS The sexual embrace of anurans in which the male's forelimbs clasp the female from behind in the region of the chest.

PECTORAL FOLD A skin fold, or groove, crossing the chest area between the forelimbs in some amphibians.

PECTORAL GIRDLE That part of the vertebrate skeleton to which the front or upper limbs are attached, consisting, in mammals, of two ventral clavicles (collar bones) and two dorsal scapulae (shoulder blades). In reptiles the clavicles are functionally replaced by the CORACOID bones.

PEDICEL *See* PEDUNCLE.

PEDICLE *See* PEDUNCLE.

PEDUNCLE; PEDICEL; PEDICLE Any small stem, or stalk, such as that produced by certain urodeles to affix their eggs to some secure object.

P GENERATION *See* PARENTAL GENERATION.

PELAGIC Of, relating to, or inhabiting, the upper waters of the open seas. Some sea snakes (genus *Pelamis*) and the marine turtles (CHELONIIDAE) are the only true pelagic reptiles, coming to land only to mate or lay their eggs.

PELOBATIDAE Spadefoot toads. Family of the ANURA, inhabiting North America, Europe, North Africa, Middle East and parts of Asia. Over 55 species in nine genera.

PELODYTIDAE Parsley frogs. Family of the ANURA (frequently placed in the PELOBATIDAE), inhabiting south-western Europe, Caucasus. Two species in a single genus.

PELOMEDUSIDAE Helmeted side-neck turtles. Family of the CHELONIA, suborder Pleurodira, inhabiting South America, Africa and Madagascar. Approximately 17 species in five genera.

PELTA A shield or a structure having the appearance of a shield.

PELTATE Shield-shaped. Sometimes used to describe the horny, epidermal LAMINA on the CARAPACE and PLASTRON of chelonians

PELVIC AMPLEXUS; INGUINAL AMPLEXUS The sexual embrace of anurans in which the male's forelimbs clasp the female from behind in the region directly in front of the hind limbs.

PELVIC GIRDLE That part of the vertebrate skeleton to which the hind or lower limbs are attached. It is absent in the majority of snakes whilst in others, and in some lizards, it is vestigial.

PELVIC GLAND Part of an enlarged, tubiform gland situated within the CLOACA in male urodeles and associated with spermatophore production. *See* CLOACAL GLAND.

PELVIC SPUR; ANAL CLAW; ANAL SPUR Either one of the two remnants of hind limbs, visible on either side of the vent in boid snakes and a few other species. In several species the spurs in the male may be noticeably longer than those of the female and can often be useful guides in the sexing of individuals.

PENIS The male copulatory organ of higher vertebrates; the INTROMITTENT ORGAN of reproduction in chelonians, crocodilians and caecilians.

PENOCLITORIS; CLITOROPENIS In examples of sex organ defect at birth, a swollen structure that could be either a small, malformed penis missing a urinary duct, or a distended clitoris. Often used in herpetology for the reproductive organ of a crocodilian which is similar in both sexes only differing in size.

PENTADACTYL Possessing five digits.

PENTADACTYL LIMB A type of limb characteristic of tetrapod vertebrates, i.e., reptiles, amphibians, mammals and birds, that has evolved from the paired fins of ancestral fishes in association with the transition from water to land. Various modifications to the basic design, consisting of an upper arm (or thigh), forearm (shank), and hand (or foot) with five fingers or toes, have been made for different functions and methods of progression, particularly by the reduction, loss or fusion of the terminal bones as adaptations to running, swimming, burrowing etc.

PERENTIE The giant monitor lizard *Varanus giganteus*, of Australia's arid interior. It reaches an average of 1.6 m in total length although specimens exceeding 2 m have been recorded making the species Australia's largest lizard.

PERINATAL Of, relating to, or occurring, directly before, during or after, birth.

PERI-ORAL Used in reference to the area around the mouth.

PERIPHERAL Any one of the marginal bones that occur as a border around the edge of the chelonian CARAPACE, each one lying beneath a MARGINAL lamina.

PERISTALSIS The rhythmic muscular contractions that occur within various bodily tubes, especially the alimentary duct, serving to transport food items and/or waste products through the body. Peristaltic contractions are probably most noticeable in snakes as they engage in swallowing large prey.

PERITONEUM A thin, translucent membrane that lines the interior of the abdominal cavity and surrounds most of the viscera. Reptiles do not possess a diaphragm and therefore lack a true abdominal cavity, having instead a COELOM. The abdominal cavity, coelomic cavity and peritoneal cavity are, for practical purposes, nevertheless synonymous.

PERITONITIS An infection of the PERITONEUM which, in reptiles, can result from a number of other conditions, such as an inability to lay eggs or deliver young (DYSTOCIA).

PERIVASCULAR HAEMORRHAGE In reference to the symptoms of certain venomous snake bites, bleeding around the blood vessels. *See* PETECHIAE.

PERMIAN The last geological period of the PALAEOZOIC era, following the CARBONIFEROUS period some 280 million years ago, and lasting some 50 million years when it was succeeded by the MESOZOIC era. The land continued to be dominated by several types of reptiles, although amphibians became reduced in both their numbers and size.

PETECHIAE (sing.) **PETECHIA** In reference to the symptoms of certain venomous SNAKE BITES, small discoloured spots on the skin's surface or mucous membrane formed from the effusion of blood following the rupture of blood vessels. Known as a 'petechial haemorrhage'.

PETECHIAL HAEMORRHAGE *See* PETECHIAE.

P GENERATION *See* FILIAL GENERATION.

pH The assessment, on a scale of 0-14, that provides a measure of the acidity or alkalinity of a medium (e.g., water). A neutral medium has a pH of 7; acidic media have pH values of less than 7 (e.g., 6-0), and alkaline media of more than 7 (e.g., 8-14). The lower the pH the more acidic the medium; the higher the pH the more alkaline. The pH of the water is often an important factor in the captive husbandry of amphibians.

PHALANX (pl.) **PHALANGES** Any one of the small bones in the digits (fingers or toes), linked to each other by hinged joints and articulating with the metacarpal bones of the fore foot or lower forelimb, or the metatarsal bones of the hind foot or lower hind limb. In the typical PENTADACTYL LIMB of five digits, there are two phalanges in the smallest and innermost digit and three in each of the others.

PHANEROGLOSSAL Possessing a tongue, as opposed to AGLOSSAL - lacking a tongue - as in, for example, members of the anuran family PIPIDAE.

PHARYNGEAL Of, relating to, or sited in or near the PHARYNX.

PHARYNX The part of the alimentary canal between the mouth and the oesophagus, serving as a passage for both food and respiratory gases. At its anterior end, it has openings from the nasal passages and mouth, and at its posterior end openings to the trachea and oesophagus. In aquatic amphibians the pharynx is pierced with gill slits through which water flows, supplying the blood in the gill filaments with oxygen.

PHEROMONE A substance that is produced and secreted by an animal and which stimulates a response in others of the same species. Pheromones have a significant role in the social behaviour of many animals and their uses include marking territories, laying trails, attracting mates etc.

PHOLIDOSIS The number, shape, pattern, arrangement and location of the scales, plates, shields and laminae of reptiles, factors important in the identification and systematics of these animals.

PHOTOPERIOD The period of light and dark (day and night) in every 24 hours. Photoperiodism, or the response of organisms to the lengthening or shortening of the diurnal photoperiod, determines the triggering of such activities as migration, hibernation, breeding and other seasonal aspects of their lives. Controlling the photoperiod of captive reptiles and amphibians is often an essential factor in successful breeding.

PHOTORECEPTOR A tissue or cell that is receptive (i.e, sensitive) to light.

PHRAGMOSIS A form of behaviour in which an animal secretes itself in a hole or small crevice and blocks the opening with its head. As well as occurring in many invertebrate species (e.g. certain wasps, ants, and spiders) phragmosis has also been observed in casque-headed tree frogs (e.g. *Corythomantis greeningi* and *Triprion petasatus*) of Central and South America. These amphibians hide in holes which they close up with their flattened bony-plated heads, giving them protection and allowing them to seal in moisture thereby slowing their rate of dehydration.

PHRYNOMERIDAE Snake-neck frogs. Family of the ANURA, suborder Diplasiocoela, inhabiting Africa south of the Sahara. Three species in a single genera.

PHRYNOSOMATIDAE Horned toads and spiny lizards etc. FAMILY of the SQUAMATA, suborder SAURIA, formerly of the IGUANIDAE. Some 125 species in

10 genera inhabiting North America from southern Canada to Panama..

PHYLOGENESIS; PHYLOGENY The entire course of development in a species, genus etc., over successive generations.

PHYLOGENETIC SYSTEMATICS The study of biological organisms, and their categorizing, based on their evolutionary descent, for the intention of classification.

PHYLUM (pl.) **PHYLA** One of the basic classificatory divisions of the animal kingdom and the highest division in TAXONOMY, comprising organisms all having the same general form. Reptiles and amphibians, together with the other invertebrates, are placed within the phylum Chordata which contains animals that possess a spinal cord.

PHYTOTHELM Any small body of rainwater occurring in a tree hollow, epiphyte funnel, leaf fold, or similar situation in forest vegetation. Important to many arboreal reptiles as a source of drinking water, and to many tree-frog species for the development of their larvae.

PIGMENT The colouring substance in the cells and tissues of animals and plants. Melanin, for example, is the pigment responsible for dark brown to black coloration in reptiles and amphibians.

PINEAL COMPLEX; EPIPYYSIAL EYE The rudimentary, median, third eye and its associated parts, in many lizards and the tuatara (*Sphenodon*), lying beneath a foramen towards the front of the parietal bone and covered only with skin and connective tissue. The complex consists of two processes projecting dorsally from the diencephalon, the anterior of which is termed the parapineal organ, and the posterior, the pineal organ, terminating at the surface on the top of the headwith the pineal eye. In the tuatara the eye is equipped with a lens, retina and nerve fibres linked to the brain, serving as a light-sensitive organ and possibly also responding to solar radiation. Derived from the pineal gland, the structure is thought to have existed in many fossil vertebrates but today its only known living example is the tuatara.

PINEAL EYE *See* PINEAL COMPLEX.

PINEAL GLAND *See* PINEAL COMPLEX.

PINEAL ORGAN *See* PINEAL COMPLEX.

PINHEAD; PINHEAD CRICKET The name given to hatchling, or extremely small, crickets.

PINHEAD CRICKET *See* PINHEAD.

PINKY Any newborn rodent that has yet to grow fur.

PINKY-PUMP A syringe-like device with a blunt tip, used in the force-feeding of snakes which, for one reason or another, will not feed voluntarily. A dead newborn rodent (PINKY) is inserted into the cylindrical barrel of the pump, the tip or nozzle of which is then introduced carefully into the snake's oesophagus. As the handle of the pump is slowly pressed into the barrel the pinky is compressed and forced out through the nozzle into the snake's stomach.

PIPIDAE Clawed frogs. Family of the ANURA, suborder Aglossa, inhabiting South America and Africa. 26 species in four genera.

PIPPING In reptiles, the act of slitting the egg by the hatchling inside, prior to its emergence. The reptile-breeder may sometimes have to pip by hand eggs which have gone POST-TERM if it is suspected that the hatchlings are too weak to do it themselves.

PISCIVOROUS Feeding on fish.

PIT; LACHRYMAL PIT; LOREAL PIT 1. The deep cavity situated on either side of the head, between the nostril and the eye, of crotaline snakes (hence the name PIT VIPER), which is the external opening to a highly sensitive infra-red detecting organ, enabling the snake to locate and strike at prey in complete darkness. *See* LABIAL PIT. **2.** A collective noun for a gathering of snakes.

PITH To kill a reptile or amphibian by piercing or severing the brain and spinal column, usually by inserting a sharp instrument between the base of the skull and the first vertebra. Often used as a form of EUTHANASIA.

PIT SCALE A scale situated in or on the margin of the PIT in crotaline snakes, such as the FOVEAL or LACUNAL scale.

PITUITARY GLAND A small, ductless gland situated at the base of the brain, the hormonal secretions of which control many important bodily functions and regulate other glands within the body.

PIT VIPER A common classification for any one of the venomous snakes in the VIPERIDAE subfamily, Crotalinae, all of which share the same characteristic feature of a sensitive, heat-detecting PIT.

PLACEBO An inactive substance with no physiological function administered to a patient who may benefit by the psychological deception.

PLACEBO EFFECT In cases of bites from non-venomous snakes in which the patient is convinced they are going to die, a positive therapeutic effect claimed by the patient after receiving a PLACEBO believed by them to be an active drug.

PLACENTA A spongy, vascular structure, uniting foetal and maternal tissues, through which the foetus receives nourishment and oxygen and its waste products are removed. The development of a placenta (placentation) is known to take place in some reptiles which exhibit VIVIPARITY.

PLACENTATION *See* PLACENTA.

PLAGIOTREME Having a transversely placed anal slit, or VENT, characteristic of species possessing paired hemipenes, e.g., snakes and lizards. *See* CYCLOTREME.

PLANTAR Of, or relating to, the sole or ventral surface (face) of the hind foot. *Contrast with* PALMAR.

PLANTAR TUBERCLE A raised, rounded, wart-like protuberance on the ventral surface (face) of the hind foot. *Contrast with* PALMAR TUBERCLE.

PLASMA The clear yellowish liquid portion of LYMPH or blood in which the cells and corpuscles are suspended.

PLASMINOGEN A protein occurring in many fluids and tissues of the body.

PLASTRAL Of, or on, or relating to, the PLASTRON of a chelonian.

PLASTRON (pl.) **PLASTRA** In reptiles of the order Chelonia (turtles, terrapins, tortoises) the ventral part of the shell consisting typically of nine symmetrically placed plate-like bones, covered on the outside by horny shields. The domed upper part of the shell is termed the CARAPACE.

PLASTRON CONCAVITY The inward-curving area of the chelonian PLASTRON, common in adult males, which helps prevent the males sliding off the females during mating.

PLATE 1. Any one of the enlarged units, formed by the fusion of smaller scales, usually on the dorsum, or upper surface, of the head, or on the under surface of the body where it is termed a VENTRAL scale. **2.** Any one of the bony elements that together form the skeleton of the chelonian CARAPACE and PLASTRON. **3.** The OSTEODERM lying beneath an epidermal scale in some reptiles.

PLATELET Any one of numerous minute disc-like particles occurring in the blood of vertebrates and serving a vital function in blood coagulation.

PLATYHELMINTHES; FLATWORMS A phylum of diverse, dorsoventrally flattened worms many of which are parasitic, some of which are well known ENDOPARASITES of reptiles and amphibians.

PLATYSTERNIDAE Big-headed turtle. Family of the CHELONIA, suborder Cryptodira, inhabiting South-East Asia. MONOTYPIC.

PLEISTOCENE The first epoch of the QUATERNARY period, beginning at the end of the PLIOCENE some 2 million years ago, and lasting until the HOLOCENE about 10 000 years ago. Characterized by a series of glacials, it is often termed the 'Ice Age' when advancing ice margins drove many organisms toward the equator, and many others to extinction.

PLETHODONTIDAE Lungless salamanders. Family of the URODELA, inhabiting North America south to central South America, with species of one genus found in southern Europe. Over 200 species in some 23 genera.

PLEURAL Any one of the bony plates forming the chelonian CARAPACE, occurring on either side of the NEURAL bones and corresponding with the COSTAL laminae.

PLEURODIROUS Descriptive of chelonians that are unable to retract the whole of their necks into the CARAPACE due to the presence of well-developed transverse processes on certain of the cervical vertebrae. The neck, which in several species is almost as long as the shell, is bent sideways beneath the front edge of the carapace. Pertaining to any member of the order Pleurodira.

PLEURODONT Possessing teeth that are situated laterally on the inner edge of the jaw bone, as in many snakes and iguanid lizards.

PLEUROGLYPH Descriptive of any rear-fanged snake (ECTOGLYPH, sense 1) in which the fang has its venom-conducting groove situated on the side. *Contrast with* PROECTOGLYPH.

PLICA (pl.) **PLICAE** Any folding over of parts, such as a muscle or fold of skin.

PLICATE Wrinkled or folded.

PLIOCENE The final epoch of the TERTIARY period, beginning at the end of the MIOCENE some 7 million years ago, and lasting about 5 million years until it was succeeded by the PLEISTOCENE.

PNEUMONIA Acute inflammation of the lungs, in which their normally spongy tissue is converted into a solid mass and the lungs themselves become filled with liquid, making them useless for breathing. In reptiles, pneumonia can be the result of bacterial or viral infections. It can also be caused by heavy infestations of the endoparasitic nematode, LUNGWORM.

POD A collective noun for a group of juvenile crocodilians.

POIKILOTHERM; ECTOTHERM Any animal in which body temperature varies approximately with that of the surroundings; a cold-blooded species. All reptiles and amphibians are poikilothermic. Contrast HOMOIOTHERM.

POISON GLAND *See* PAROTOID GLAND; VENOM GLAND.

POISONOUS In general, used to describe any living organism having a toxic secretion, or a bite or sting which discharges a poisonous substance, acid or venom, a comparatively small quantity of which can cause sickness or death. It is, however, becoming more acceptable to restrict the term to organisms that have a harmful effect on another when partially or completely devoured, and to use the term VENOMOUS for organisms that introduce venom into the body of another by means of a sting or, as in certain reptiles, the teeth.

POLLEX (pl.) **POLLICES** The first, or innermost, digit on the forelimb of a tetrapod vertebrate, containing two phalanges.

POLLIWOG A name used in some regions for a newly-metamorphosed frog.

POLLYX A variant spelling of POLLEX.

POLYCHROTIDAE Anoles. Family of the SQUAMATA, suborder SAURIA, formerly of the IGUANIDAE, inhabiting the Caribbean and Central and South America with over 370 species in some four genera.

POLYDONT Possessing a number of fully functional teeth.

POLYLYMPHADENOPATHY
Having multiple enlarged lymph
nodes.

POLYMORPHIC *See*
POLYMORPHISM.

POLYMORPHISM The occurrence
of more than two types of individual
within a species or subspecies,
determined genetically or by the
environment. Several reptiles (e.g.,
California king snake, *Lampropeltis
getula californiae*) and amphibians
(e.g., African reed frog, *Hyperolius
marmoratus*) are polymorphic, all
having more than two pattern or
colour types.

POLYPHYODONT Having teeth
that are replaced more than once
during the animal's life.

POLYTYPIC Containing several
forms (TAXA) e.g., a genus with
more than two species. *Contrast*
MONOTYPIC.

POLYVALENT Possessing
antibodies for more than a single
particular organism or substance, and
a term used commonly for any
antivenom effective against the
venoms of a number of different
snake species.

POPE, Clifford H., (1899-1974),
American herpetologist. On
graduating from the University of
Virginia he joined the Department of
Amphibians and Reptiles at the
American Museum of Natural
History and, after some five years as
a field naturalist with the Central
Asiatic Expeditions in China, he was
made Assistant Curator of
Herpetology at the Museum. Joined
the Chicago Natural History
Museum in 1940 where he became
Curator of Reptiles and Amphibians
until 1953. His publications include
Reptiles of China (1935), *Turtles of
the United States and Canada*
(1939), *Snakes alive and how they
live* (1942), *Amphibians and reptiles
of the Chicago area* (1944), *Giant
snakes* (1962), and *The reptile world*
(1955).

POPPING The term used to describe
the method of sexing juvenile
snakes, involving the application of
gentle pressure to the base of the tail
with the ball of the thumb which is
slowly rolled upwards towards the
CLOACA. In most species, if the
snake is a male it will event, or
'pop', its HEMIPENES.

POPULATION 1. A collection of
individuals of the same species
occupying a particular space. A
population is determined by birth and
death rates, male/female ratios,
density etc., and is controlled by
various environmental factors,
disease and food supply etc. 2. The
total number of individuals of a
given species occurring within a
defined area, e.g., the population of
the smooth snake *Coronella
austriaca* in Britain is estimated at
around 2000 adults.

POPULATION ECOLOGY The
study of the relationship of a specific
species or genus (or higher taxon)
with its surroundings.

PORE Any small opening or passage
in the skin or external covering as,
for example, the FEMORAL and
PREANAL PORES of lizards which
may, in breeding males, secrete a

158

wax-like substance.

POST- A prefix meaning following; behind, subsequent.

POSTABDOMINAL The enlarged ANAL scale, or plate, of snakes.

POSTANAL Behind, or posterior to, the anus or cloacal region.

POSTANAL GLAND The ANAL GLAND of snakes, often more prominent in the females.

POSTANAL SAC Either one of the pair of pouch-like hollows situated directly behind the vent on either side of the tail in some lizard families, e.g., ANGUIDAE, GEKKONIDAE and PYGOPODIDAE, and thought to be similar in function to the CLOACAL GLAND of crocodilians and the ANAL GLAND of snakes.

POSTANAL SCALE In the males of some lizard species, a scale lying posterior to the anus which, in iguanids, is usually enlarged and paired.

POSTANAL TRIANGLE An often sharply contrasting area of ventral scales in male EUBLEPHARID lizards, the apex of which is in the area of the vent, with the sides of the triangle diverging along the rear margin of the thighs. Also known as the ESCUTCHEON.

POSTANAL TUBERCLE A seldom used alternative term for the PELVIC or ANAL SPUR.

POSTAURICULAR Referring to the area of the neck behind the ear. Postauricular swellings frequently occur in certain lizards (e.g., *Phelsuma*) especially in reproductively active females, and

are thought to play an important role in the metabolism of calcium during egg-shell formation.

POSTAXIAL Of, or relating to, the posterior side of the forelimb.

POSTCENTRAL The single or paired SUPRACAUDAL lamina on the rear edge of the CARAPACE in chelonians.

POSTERIOR Located at, or towards, the rear of, or near, the back. *Compare* ANTERIOR.

POSTERIOR LOBE The section, lying posteriorly to the INGUINAL NOTCH, of a chelonian PLASTRON.

POSTFOVIAL Any one of the small scales bordering the PIT of crotaline snakes, lying between the LACUNAL and INTEROCULABIAL scales. *See* FOVEAL; PREFOVEAL.

POSTFRONTAL Any one of the scales situated on the head of a worm lizard (AMPHISBAENIDAE).

POSTGENIAL In worm lizards (AMPHISBAENIDAE), a scale situated beneath the chin.

POSTLABIAL In lizards, one or more enlarged scales, or plates, situated behind the LABIAL scales.

POSTLOREAL In rattlesnakes of the genus *Crotalus*, any small scale occupying the space between the LOREAL and PREOCULAR scales.

POSTMALAR Any one of a series of chinscales in the South American worm lizards (*Amphisbaena*) posterior to the CHINSHIELD and the MALAR.

POSTMANDIBULAR Any one of the scales forming a series directly beneath and to the rear of the lower labials in chelonians.

POSTMARGINAL The SUPRACAUDAL lamina of a chelonian CARAPACE.

POSTMAXILLARY Pertaining to teeth situated in the upper posterior part of the jaw; posterior to the fangs.

POSTMENTAL In snakes the single, and in lizards the single, or paired, scale situated directly posterior to the MENTAL scale.

POSTNASAL The scale, or scales, situated in front of the LOREAL and behind the NASAL scale.

POSTNEURAL An alternative term for the bony SUPRAPYGAL plate on the posterior of a chelonian carapace.

POSTOCULAR Behind the eye, as in 'postocular scale' which refers to any scale situated on the rear edge of the eye socket (ORBIT).

POST-PARTUM; **POST-PARTURIENT** The period of time following birth (PARTURITION).

POST-PARTURIENT *See* POST-PARTUM.

POSTROSTRAL In some snake species, e.g., rattlesnakes (*Crotalus, Sistrurus*), any scale lying between the ROSTRAL and the INTERNASAL scales.

POSTSYNAPTIC MEMBRANE The membrane of a muscle which, in response to a nerve impulse, is stimulated to begin contraction. *See* PRESYNAPTIC MEMBRANE.

POSTSYNAPTIC NEUROTOXIN A NEUROTOXIN which works on the muscle side of the synaptic gap and causes a form of paralysis which is more easily reversed using antivenom than that caused by a PRESYNAPTIC NEUROTOXIN. The venom of the death adders *Acanthophis* spp. and common cobras *Naja* spp. are post-synaptic.

POST-TERM After term; after the due date. Said of eggs which have gone beyond the date they were expected to hatch.

POUCH *See* BROOD POUCH; MARSUPIAL POUCH.

POX; LIZARD POX A harmless dermal condition of unknown origin, occurring occasionally in both wild and captive lizards, especially *Lacerta*, in which one or more tumours (papillomata) form in the skin. These tumours sometimes develop into enlarged, roughened and grotesquely shaped structures known commonly as 'tree bark tumours'.

PRE- A prefix meaning before in time, rank, position, order etc.

PREANAL In front of the anus or cloacal region. *See* ANAL (sense 2).

PREANAL PORE A structure, identical in function to the FEMORAL PORE, situated anterior to the anus or cloacal region, in some lizards. The pore opens to the exterior on the upper surface of the PREANAL SCALE.

PREANAL SCALE Any one of the row of scales situated in the pelvic region, directly in front of the anus. In some lizards several of these scales may have PREANAL PORES.

PREBUTTON In newborn rattlesnakes (*Crotalus, Sistrurus*) the small pre-rattle segment on the tip of the tail which is usually lost a short while after birth when the snakes undergo their first skin shedding.

PRECAMBRIAN The period of geological time extending from the earth's formation, considered to be about 4600 million years ago, to the beginning of the CAMBRIAN some 590 million years ago.

PRECENTRAL The NUCHAL LAMINA of the chelonian CARAPACE.

PRECILIARY In rattlesnakes (*Crotalus, Sistrurus*), a small scale occupying the front upper border of the eye.

PREDATION A relationship between two or more SPECIES within a community, in which one (the predator) hunts the other (the prey), which may be vertebrate or invertebrate, for food. Most reptiles, and amphibians and their larvae, are predators.

PREFERRED BODY TEMPERATURE The temperature at which a reptile feels comfortable and able to remain active in order to maintain its regular activities such as feeding, basking, courting and mating etc.

PREFERRED SUBSTRATE TEMPERATURE The temperature of the ground upon, or in which, a reptile chooses to rest, when the substrate of that particular area has a varying thermal gradient.

PREFOVEAL Any one of the small scales in crotaline snakes (pit vipers) bordered by the NASAL, LOREAL, SUPRALABIAL and LACUNAL scales. *See* FOVEAL; POSTFOVEAL.

PREFRONTAL Either one of a pair of large scales directly in front of the FRONTAL.

PREGULAR FOLD Any one of the single, or several, skin folds transversing the throat in some lizards, in front of the GULAR FOLD.

PREHALLUX The vestigial digit on the innermost side of the first toe (HALLUX) of the hind foot in anurans.

PREHENSILE Adapted for seizing or grasping, especially by wrapping around a support. The tail of the nocturnal monkey-tailed skink *Corucia zebrata* is prehensile and used for grasping branches from which it hangs during the day whilst asleep.

PRE-MATING ISOLATING MECHANISM The term given to the unique individual CALL of anuran species to which only other members of the same species will respond. Where several species occur together within the same habitat, this auricular 'signature' is vital as it prevents females of one species from mating with males of another, thereby saving valuable energy and time on a sterile mating. The call thus effectively 'isolates' one species from another.

PREMAXILLA BONE Either one of the two, often teeth-bearing, dermal bones that form the front-most angle of the upper jaw in amphibians and reptiles.

PREMAXILLA TOOTH Any one of the teeth found on the PREMAXILLA BONES of the front upper jaw.

PREMONTANE Said of reptile and amphibian species that occur in the foothills of a mountain or mountain range.

PRENASAL A scale situated directly in front of the nostril in some snakes.

PREOCCUPIED In accordance with the code of zoological nomenclature, the rule of priority in which the first name has precedence and cannot be used again for another species.

PREOCULAR; ANTEOCULAR; ANTEORBITAL Situated in front of the eye, as in 'preocular scale' which refers to any scale situated on the front edge of the eye socket (ORBIT).

PRESSURE RIDGE

PRESUBOCULAR Any one of the several small scales situated beneath the PREOCULAR and in line with the LOREAL. Present in the dharman ratsnake *Ptyas mucosus* and Indo-Chinese ratsnake *Ptyas korros*

PRESYNAPTIC MEMBRANE The membrane of a nerve ending from which a neurotransmitter passes into SYNAPSE en route to the postsynapse.

PRESYNAPTIC NEUROTOXIN A NEUROTOXIN which works on the nerve side of the synaptic gap and causes a form of paralysis which is more difficult to reverse using antivenom than that caused by a POSTSYNAPTIC NEUROTOXIN. The venom of the taipans *Oxyuranus*, the blacksnakes *Pseudechis* and the brownsnakes *Pseudonaja* are pre-synaptic.

PRE-TERM Before term; before the due date. Said of eggs that hatch earlier than expected.

PRETYMPANIC; PRETYMPANUM Any one of the number of scales situated directly in front of the tympanum, or ear, on either side of the head in lizards.

PREVERNAL Pertaining to the early spring. *Compare* AUTUMNAL, HIBERNAL, SEROTINAL, VERNAL.

PREY Any animal that is hunted or captured by another animal for food. Even though the majority of both reptiles and amphibians are predators, most of them are also, at some stage in their life, prey not only to other forms of animals but frequently to their own kind as well. *See* PREDATION.

PREY SPECIES SPECIFIC In reference to feeding, the situation in which a particular food item is eaten in preference to any other. Reptiles that use a specific prey species exclusively are termed 'stenophagous'; the egg-eating snake *Dasypeltis scabra* is an example.

PROBOSCIS; ROSTRAL APPENDAGE The long tubular snout, or ROSTRUM, in the soft-shelled turtles (*Trionyx*), matamata *Chelus fimbriatus* and Fly River turtle *Carettochelys insculpta*, which can be protruded above the water's surface like a snorkel, allowing the

162

chelonian to breath whilst the rest of the head and body remains submerged.

PROCAUDAL *See* SUPRAPYGAL.

PROCESSUS ARTICULAR *See* ZYGAPOPHYSIS.

PROCOAGULANT *See* COAGULANT.

PROCRYPTIC Possessing camouflaging patterns or colours; exhibited in many reptiles and amphibians and helping them to remain undiscovered amongst the surroundings in which they rest.

PROECTOGLYPH Descriptive of any rear-fanged snake (ECTOGLYPH) in which the fang has its venom-conducting groove situated on its front surface. *Contrast with* PLEUROGLYPH.

PROGENESIS A type of PAEDOMORPHOSIS in which growth is arrested by the premature arrival of maturity, with the resulting adult specimen resembling exactly a juvenile stage of its ancestor.

PROLAPSE; PROLAPSUS The sinking of an organ, or part, from its normal position within the body. In amphibians, it can occur as a rectal and cloacal prolapse, the causes usually being constipation and/or a vitamin or mineral deficiency, and is not uncommon in *Ambystoma* salamanders and giant anurans. In reptiles, it usually occurs as a prolapse of the oviduct, through EGG-BINDING, or the rectum, through constipation.

PROLONGED AMPLEXUS The system in some anurans in which the male and female remain coupled for days, weeks or, as in certain members of the *Atelopus* genus, several months, whilst they travel vast distances necessary to reach their breeding sites.

PRONEURAL; NUCHAL BONE The central bony plate at the front edge of a chelonian CARAPACE, corresponding to the NUCHAL, or PRECENTRAL LAMINA.

PROPHYLACTIC *See* PROPHYLAXIS.

PROPHYLAXIS Precautionary treatments applied in order to prevent disease or contain its spread. It is often advisable to give prophylactic treatment to any reptile or amphibian that is suspected of carrying disease, or disease-carrying agents, or to newly imported or wild-caught specimens before introducing them to a collection.

PROTANDRY In herpetology, the arrival of males before females at breeding sites as, for example, in various reptile and amphibian species in which the males develop sperm before the females develop eggs. *Compare* PROTOGYNY.

PROTEIDAE Waterdogs, mudpuppies and olms. Family of the URODELA, inhabiting North America and Europe. Some five species in two genera.

PROTEINURIA In reference to the symptoms of certain venomous snake bites, the presence of protein, usually ALBUMIN, in the urine.

PROTEOLYTIC *See* CYTOTOXIN.

PROTEROGLYPH A snake that possesses venom-conducting fangs, usually with a closed groove, connected to a comparatively immobile MAXILLA. Consequently the fangs of proteroglyphous (or proterglyphic) snakes cannot be folded back against the roof of the mouth. *Contrast* SOLENOGLYPH.

**PROTEROGLYPHIC;
PROTEROGLYPHOUS** *See* PROTEROGLYPH.

PROTHROMBIN The constituent of blood plasma that aids the formation of clots.

PROTOGYNY In herpetology, the arrival of females before males at breeding sites, or the condition in which the females develop eggs before the males produce sperm. *Compare* PROTANDRY.

PROTOZOAN Any minute, single-celled invertebrate of the phylum Protozoa, including flagellates, amoebas, ciliates, sporozoans etc., some of which occur as ENDOPARASITES in reptiles and amphibians.

PROTOZOONOSIS Any disease of the digestive tract or blood cells and circulatory system, in reptiles and amphibians, caused by an infestation of Protozoa.

PROVENANCE The history of an individual specimen, detailing its origin, its parents, diseases and treatments, breeding successes etc.

PROXIMAL Pertaining to the part of a structure or organ, such as the tail or limb, that is closest to its point of attachment to the body. *Contrast* DISTAL.

PSAMMOPHILOUS Of, or relating to, or living in, a sandy substrate.

Several amphibians and many reptiles are psammophilous in their habits.

PSEUDIDAE Harlequin frogs. Family of the ANURA, inhabiting the Caribbean coast of Venezuela and Trinidad down to Patagonia, southern Brazil, Uraguay, Paraguay and northern Argentina. Four species in two genera.

PSEUDOMONAD INFECTION An infectious disease affecting the digestive tract, oral cavity, stomach and tissues of reptiles caused by various bacteria such as *Pseudomonas fluorescens* and *Pseudomonas aeruginosa*. MOUTH ROT in snakes is a relatively common pseudomonad infection.

PSEUDOTUMOUR An abnormal swelling in the skin of reptiles and amphibians, caused either by encysting parasites, such as nematode worms, or by an increase in the number of cells (hypertrophy) which may occur in the course of the healing of a wound or during REGENERATION.

PST Abbreviation for PREFERRED SUBSTRATE TEMPERATURE.

PTERYGOID Either one of the pair of bones that form much of the palate in many species, secured in front to the upper jaw and roof of the mouth and at the back to the QUADRATE. They are supported, in some species, by the two vertical, strut-like EPIPTERYGOID BONES.

PTERYGO-PALATINE TEETH Teeth situated behind the PALENTINE TEETH.

PTOSIS In reference to the

symptoms of certain venomous SNAKE BITES, the drooping of the eyelids resulting from paralysis of the muscles which usually raise them. Ptosis is one of the first symptoms indicative of NEUROTOXIC envenomation.

PUBIC SYMPHYSIS A joint created from the fusion of the two pubic bones (PUBES) of the pelvic girdle in many reptiles.

PUBIS (pl.) **PUBES** One of the three bones that constitute part of the PELVIC GIRDLE. The pubes in many reptiles are fused together forming a joint with a small degree of movement, the PUBIC SYMPHYSIS.

PULMONARY Of, or pertaining to the lungs.

PULMONARY OEDEMA In reference to the symptoms of certain venomous snake bites, the escape of fluids into the ALVEOLAR spaces of the lungs.

PUPIL The opening (aperture) in the centre of the IRIS of the eye. In reptiles and amphibians, the pupil can be any one of a number of shapes: circular, vertically elliptical, horizontally elliptical, heart-shaped or triangular, all useful features in the identification of these animals. The size of the pupil can be adjusted by the contraction of the muscles of the iris in order to protect the sensitive cells of the retina.

PUTREFACTION *See* PUTREFACTIVE.

PUTREFACTIVE Resulting in the decomposition or putrefaction, with an offensive smell, of animal tissue.

PVA Abbreviation for POPULATION VIABILITY ANALYSIS.

PYGAL The central bony plate at the rear edge of a chelonian CARAPACE, corresponding to the SUPRACAUDAL lamina.

PYGOPODIDAE Flap-footed, or snake, lizards. Family of the SQUAMATA, suborder SAURIA, inhabiting Australia and New Guinea and a few adjacent islands including New Britain. 31 species in eight genera.

PYREXIA In reference to the symptoms of certain venomous snake bites, having a fever in which the body temperature is above 37°C.

Q

Q_{10} A measure of the amount of oxygen used by an animal for every 10°C increase in the ambient temperature.

QUADRAT In ECOLOGY, a square sample area, which can be any size, chosen for detailed study of, for example, the amphibian or reptile POPULATION of a square kilometre.

QUADRATE 1. A large scale situated on the head, forming the hindmost end of the JUGAL group. **2**. A bone of the skull, forming the jointed lower jaw surface, more conspicuous in reptiles than amphibians in which it is usually small or absent altogether.

QUADRATOJUGAL A dermal bone situated on the BUCCAL area beneath the SQUAMOSAL, and between the JUGAL and the QUADRATE on the lower edge of the skull. Found in crocodiles, chelonians and *Sphenodon* but not lizards and snakes.

QUADRATOMAXILLARY In anurans, a bone situated between the QUADRATE and the MAXILLA, in the skull.

QUADRICARINATE Having four ridges, or KEELS.

QUADRIFURCATE Having four spines or branches of equal length.

QUADRUPEDAL Descriptive of animals that walk on four legs.

QUARANTINE A period, or place, of isolation for newly acquired or imported specimens in order to prevent the spread of disease or parasites etc. For reptiles and amphibians, the period of quarantine is generally four to six weeks, during which time the animals are observed for signs of ill-health and treated accordingly. A preventive drug (PROPHYLAXIS) is frequently given to specimens suspected of having, but not showing signs of, parasites or illness.

QUATERNARY The second, and most recent, period of the CENOZOIC from around 2 million years ago to the present, and consisting of two epochs: the PLEISTOCENE and RECENT. During the Quaternary, or 'fourth age', man became the dominant terrestrial species.

QUINCUNX A group of five objects arranged in a rectangle or square with one object at each of the four corners and a fifth in the centre. Has been used to describe the laminae arrangement on the CARAPACE of *Chrysemys*.

QUINQUECARINATE Has been used to describe laminae or scutes bearing five KEELS.

QUINTANGULAR Having five angles.

QUIVER A collective noun for a group of cobras.

qv. Abbreviation denoting a cross-reference.

R

RACE A group, or population, of a species distinguished from other members of the same species by different physical characteristics, such as colour and markings, and inhabiting a more or less isolated geographical area; a SUBSPECIES.

RACER Any one of a number of long, slender, non-venomous snakes of the family COLUBRIDAE, subfamily Colubrinae, genus *Coluber,* inhabiting North America, Europe, north-western Africa, the Near East and central Asia. In Europe they are usually known as 'whip snakes'.

RADIOGRAPHIC DIAGNOSIS The use of X-rays to confirm the presence of eggs or young, or to determine the nature of an intestinal obstruction or abdominal tumour, in a reptile.

RADIO TRACKING An observational system used to study animal behaviour in the field, in which the animal is fitted with a radio transmitter, generally attached to a collar, and the observer is equipped with a receiver and directional antenna.

RADIOULNA The bone of the lower forelimb in anurans, formed from the fusion of the RADIUS and ULNA.

RADIUS The shorter of the two bones of the lower forelimb (forearm) in tetrapod vertebrates. In anurans it is fused with the ULNA to form the RADIOULNA.

RAINFOREST A humid, evergreen forest with thick and lush vegetation characterized by huge trees, lianas, palms, epiphytes etc., and with a tremendous diversity of reptile and amphibian species. Rainforests occur mainly in lowlands of the tropical wet zone.

RAIN SHADOW The dry region on the leeward side of a mountain or mountain range, in which the annual rainfall is markedly less than that on the windward side. For example the White Mountains of east central California are in the rain shadow of the Sierra Nevada.

RAKER The GILL RAKER of larval amphibians.

RALE The term given to the abnormal crackling sound, heard through a stethoscope, when listening to lungs which have an accumulation of fluid, and frequently heard in snakes suffering from respiratory infections, pneumonia etc. *See* RHONCHUS.

RAMUS (pl.) **RAMI** Any part, organ or process that branches from another. One half of the lower jaw or MANDIBLE.

RANAVIRUS A genus of the Iridoviridae virus family able, in amphibians, to instigate diseases with a high death rate. Several ranaviruses can also infect fish and reptiles.

RANIDAE True frogs. Family of the ANURA, inhabiting much of the world with the exception of Australia, New Zealand, the southernmost part of South America and small islands. 667 species in 47 genera.

RAPHE An elongated, seam-like ridge along the union of the two halves of an organ, or part of the body, such as that down the centre of the back in certain anurans along which the outer layer of skin (STRATUM CORNEUM) breaks at the start of shedding (ECDYSIS).

RASPING ORGAN The arrangement of rows of numerous minute rasping cusps around the mouth in the majority of anuran tadpoles, with which they scrape particles of food from submerged rocks and other debris.

RATTLE; RATTLE-STRING; STRING The sound-producing organ on the tails of rattlesnakes (*Crotalus, Sistrurus*), formed from a series of keratinized rings or SEGMENTS, each stacked loosely within the other, a new one being added at each skin shedding. Responsible for the characteristic rattling or buzzing noise when the tail is vibrated during moments of excitement.

RATTLE-FRINGE The posterior caudal scales of a rattlesnake (*Crotalus, Sistrurus*) which extend over the edge of the basal segment of the RATTLE.

RATTLE-STRING An alternative term for a rattlesnake's RATTLE.

RAW NOSE The general term given to the common condition often seen in a particularly nervous or overactive captive reptile, especially one housed in an enclosure which is too small, in which rostral damage is caused to its snout. It is a result of the lizard or snake continually rubbing its snout against the glazed surfaces of its enclosure in its attempt to escape or reach things it can see on the other side of the glass. The rostral scale on the tip of the snout is consequently eroded and bleeding occurs. Although any bleeding will cease once the reptiles settles down, unless something is done swiftly to halt the behaviour the scale, along with some or all of the adjoining scales and possibly underlying bone, will be permanently lost and scarring will remain.

RAY The inner bony structure of the fin-like crest on the body and tail of certain lizard species e.g., *Basiliscus*, formed by extensions on the dorsal vertebrae.

RAY, John, (1627-1705), British anatomist and naturalist. His generalized account of reptiles, *Synopsis methodica animalium quadrupedum et serpentini generis,* in 1693, comprehensive and excellent for its time, distinguished the heart of reptiles from that of mammals and birds, and was the first to precisely describe the tooth structure in venomous and non-venomous snakes, showing that it was possible to distinguish venomous from harmless species by their dental characteristics.

REAR-FANGED; BACK-FANGED; OPISTHOGLYPHOUS Possessing enlarged, grooved, venom-conducting teeth situated in the rear of the mouth on the upper jaw (MAXILLA). Rear-fanged snakes are found in the family COLUBRIDAE, in several genera including *Boiga, Dispholidus, Ahaetulla* and *Thelotornis*, among others.

RECENT; HOLOCENE The present epoch in the geological time scale, and the second of the QUATERNARY period, extending from the last glaciation of the PLEISTOCENE, some 10 000 years ago, to the present day.

RECEPTACULUM SEMINIS *See* SPERMATHECA.

RECRUDESCENCE Reappearance or regrowth. Used in reference to testicular cycles in male reptiles, following a REFRACTORY PERIOD.

RECTIFORM; LONGITUDINAL TYPE Descriptive of a particular arrangement in the scalation of snakes; a SCALE ROW forming a straight rather than an oblique series.

RECTILINEAR MOVEMENT; CATERPILLAR MOVEMENT The method of progress in snakes, in which the reptile proceeds forwards in a straight line by means of complex alternate movements of the ribs, muscles and large ventral scales, in a series of undulations which involve no lateral movement.

RECURVED Descriptive of a tooth that bows backwards.

RED ALBINO A term used occasionally for AMELANISTIC specimens.

REDBUG *See* CHIGGER.

RED-LEG One of the most dangerous bacterial diseases affecting captive aquatic and semi-aquatic amphibians, easily and rapidly spread from one animal to another by the bacterium *Aeromonas hydrophila*, and resulting in a distinct reddening of the under surfaces of the hind limbs and, in some cases, lower abdomen; a form of severe generalized haemolytic sepsis, from which most victims fail to recover.

REFLEX FEEDING Pertaining usually to newborn or freshly captured snakes which are reluctant to feed voluntarily and are induced to do so by encouraging them to use their strike reflex, accomplished by provoking them into striking at, seizing and, subsequently, swallowing the item of food.

REFRACTORY PERIOD In reptiles, the period, usually lasting for several months following mating, in which testicular regression may take place in males, with the production of sperm only recurring again after exposure to a period of cooling followed by a gradual warming. Ovulation in females may also be reduced or halted during this period.

REFUGE Any place, within or beneath which a reptile or amphibian can find shelter and seclusion from cold, heat, light, predators and others of its own kind etc. Such refuges may be nothing more than a rock crevice; a hollow log; a piece of corrugated iron; a compost heap; or, in the case of captive specimens, a hide box. Not to be confused with REFUGIUM.

REFUGIUM (pl.) **REFUGIA** Any geographical region that has remained unaltered by a climatic change affecting surrounding regions and that consequently forms a safe haven for relict species.

REGENERATION The re-growth, in a reptile or amphibian, of an organ lost or damaged through injury, AUTOTOMY etc. Urodeles can replace part of, or an entire, limb with an identical one, and lost tails in many reptile species can be re-grown.

RELICT A group of organisms surviving as a remnant of a vanishing, formerly widely distributed race, type or species, usually in an environment different from that in which it originated; of an almost extinct order such as, for example, the tuatara *Sphenodon punctatus* and the Bornean earless monitor *Lanthanotus borneensis*.

RENAL Of, relating to, resembling, or located near the kidney.

RENIFORM Having the shape or outline of a kidney, as, for example, the PAROTOID GLAND of some bufonid toads.

REPLACEMENT FANG; ACCESSORY FANG; RESERVE FANG In front-fanged venomous snakes, any one of the teeth that lie, in progressive stages of development, within a protective fleshy sheath behind the functioning fang, the next in succession and most developed of which serves as a replacement should the original be lost.

REPRODUCTIVE ISOLATION Any situation in which different species living within the same area will not interbreed as a result of some behavioural or biological factor.

REPRODUCTIVE PERIOD The interval in time during which males and females engage in sexual activity. In temperate reptile and amphibian species this may be a defined period in the spring following HIBERNATION. In tropical species, which experience a largely stable climate, the period can occur at any time during the year.

REPTILIA Reptiles. The class of vertebrates that contains the first entirely terrestrial tetrapods, and which consists of the crocodilies and alligators (CROCODILIA), snakes and lizards (SQUAMATA), tuatara (RHYNCHOCEPHALIA), tortoises and turtles (CHELONIA), and the amphisbaenids which form a suborder of the Squamata. Known since the Upper Carboniferous, they number today approximately 6000 species. Most possess a dry skin with a covering of keratinized scales or shields preventing water loss. All are POIKILOTHERMS, and respiration is by lungs only. Fertilization is internal and there is no larval stage or METAMORPHOSIS.

REPTILIARY An open-air enclosure in which reptiles are maintained in near natural conditions.

RESERVE FANG *See* REPLACEMENT FANG.

RESONATING ORGAN The VOCAL SAC of male anurans, situated on the throat or on the sides of the neck.

RESORPTION; RESORBTION In reptilian reproduction, the act of the female in re-absorbing fully developed eggs. The cause of resorption in female reptiles may be due to unfavourable environmental conditions or physical abnormalities but because the exact reason is presently unknown there is no recognized method of prompting or averting it.

RESOURCE PARTITIONING The nature by which species living within the same area share the same resources of territory, food, and activity periods etc.

RESPIRATION The process in living organisms of taking in oxygen and giving out carbon dioxide. In reptiles this is achieved by breathing air through the lungs, whilst in amphibians it is effected in several ways: in most adult terrestrial forms, via lungs and, in adult aquatic forms and most larval stages, via gills. All adult and larval amphibians also absorb oxygen through their skin from the air or water around them.

RESPIRATORY DISTRESS In reference to the symptoms of certain venomous snake bites, difficulty in breathing.

RETICULATE Netlike in form or pattern; a skin pattern consisting of a network of intersecting lines and blotches, as in the reticulated python *Python reticulatus*, for example.

RETINA The light-sensitive, membranous layer lining the rear wall of the eyeball, and consisting of cones (most numerous in diurnal reptiles and amphibians) and rods (most numerous in nocturnal ones), which are sensitive to colour and light. Images focused on the retina are transmitted, as nerve impulses, to the brain.

RETRACTILE Capable of being drawn inwards, e.g., the head and limbs of many chelonians can be retracted into the SHELL.

RHABDIAS INFECTION Disease caused by the parasitic NEMATODE of the genus *Rhabdias,* infecting the respiratory system and other organs of anurans. The larvae of *Rhabdias sphaerocephala* have been observed to quickly work their way out through the skin of the cane toad *Bufo marinus* resulting in overnight death. *Rhabdias* nematode species reach the lungs either by direct migration or indirectly via the blood stream but can encyst other organs such as the eye, liver and heart etc.

RHABDOMYOLYSIS *See* RHABDOMYOLYTIC.

RHABDOMYOLYTIC In reference to certain venomous SNAKE BITES, said of a venom which strikes skeletal muscle resulting in the loss of muscle protein (myoglobin) into the urine, and the possibility of subsequent kidney failure. Certain seasnakes have venoms which are predominantly rhabdomyolytic and the venoms of other species may also cause rhabdomyolysis to a lesser degree.

RHACOPHORIDAE Flying frogs. Family of the ANURA, inhabiting Africa, Madagascar and the Oriental region. Over 200 species in 10 genera.

RHAMPHOTHECA The sharp, keratinous sheath, or BEAK, on the jaws of a chelonian.

RHEOCOLOUS Of, situated in, or inhabiting, the waters of streams. Many urodeles are rheocolous, e.g., the spring salamander *Gyrinophilus porphyriticus* and brook salamander *Eurycea bislineata*, among others.

RHINEURIDAE 'Worm lizards'. Family of the SQUAMATA, suborder Amphisbaenia, inhabiting central and northern Florida. MONOTYPIC genus.

RHINODERMATIDAE Mouth-brooding frogs. Family of the ANURA, suborder Procoela, inhabiting southern Chile and Argentina. Two species in a single genus.

RHINOPHRYNIDAE Mexican burrowing toad. Family of the ANURA, suborder Opisthocoela, inhabiting Central America from Texas to Costa Rica. MONOTYPIC genus.

RHOMB A lozenge- or diamond-shaped marking in the skin patterns of snakes, e.g., the rhombic night adder *Causus rhombeatus*.

RHOMBOID Lozenge- or diamond-shaped.

RHONCHUS (pl.) **RHONCHI** The term given to the abnormal snapping sound, heard through a stethoscope, when listening to lungs which have an accumulation of fluid, and frequently heard in snakes suffering from respiratory infections, pneumonia etc. *See* RALE.

RHUMBA A collective term for a gathering of rattlesnakes.

RHYNCHOCEPHALIA; 'BEAK-HEADS' The order of primitive, superficially lizard-like reptiles of the order LEPIDOSAURIA, suborder Rhynchosaurida (giant beaked lizards), common during the MESOZOIC but now extinct except for one living species, the tuatara *Sphenodon*, characterized by (among other features) a DIAPSID skull, a well-developed PINEAL EYE and a beak-like, toothless upper jaw.
See SPHENODONTIDAE.

RICKETS *See* NUTRITIONAL SECONDARY HYPERPARA-THYROIDISM.

RICTUS The gape, or mouth opening.

RING 1. A colour pattern element consisting of a broad, or narrow, solid band of colour that completely encircles the long axis of the body of a snake. **2.** Any one of the individual, loosely connected keratinized segments that constitute the RATTLE in rattlesnakes (*Crotalus, Sitrurus*).

RINGHALS; RINKHALS The much feared spitting cobra *Hemachatus haemachatus* of the family ELAPIDAE, subfamily Elapinae, inhabiting Africa south of the Zambezi.

RIPARIAN; RIVERINE Of, relating to, or inhabiting, the bank of a river, lake or other body of water.

172

RITUAL COMBAT Intraspecific and innate rivalry in male reptiles and amphibians during the breeding season involving two or more participants, in which each attempts to overthrow the other in often elaborate, but rarely injurious fights, usually in order to gain possession of a female.

RIVERINE *See* RIPARIAN.

RIVER JACK The rhinoceros viper *Bitis nasicornis*, a member of the family VIPERIDAE, subfamily Viperinae, inhabiting rainforests of West Africa east to Uganda, western Kenya and southern Sudan, and southwards to Angola.

RNS Abbreviation for RUNNY NOSE SYNDROME.

ROMER, **Alfred Sherwood**, (1881-1950), American palaeontologist and anatomist, Director Emeritus of the Museum of Comparative Zoology at Harvard University. Author of several publications dealing with the anatomy of reptiles, including *Osteology of the reptiles* (1956), containing a detailed account of the skeleton of both fossil and modern forms, and *Vertebrate palaeontology* (1945, 1966).

ROSTRAL; APICAL Pertaining to the ROSTRUM (snout), e.g., the rostral scale situated on the tip of the snout between the LABIAL scales that border the upper lip.

ROSTRAL APPENDAGE The long tubular snout (PROBOSCIS) in certain chelonians (e.g., soft-shelled turtles, *Trionyx*).

ROSTRAL CREASE A distinct groove or indentation (the

LINGUAL FOSSA) in the rostral scale of snakes, allowing the protrusion of the tongue when the mouth is closed.

ROSTRAL HUMP The enlarged scaly hump on the tip of the snout of the lyre-headed agamid lizard *Lyriocephalus scutatus*.

ROSTRUM (pl.) **ROSTRA** The BEAK, or a beak-like part. Used in herpetology for the snout of certain reptiles.

ROSY A recently devised term for a variety of corn snake *Pantherophis guttatus guttatus* found in the lower Florida Keys and displaying a minimal amount of black coloration both dorsally and ventrally.

ROUND WINDOW *See* TYMPANUM.

ROUNDWORM Any worm of the NEMATODA, some of which are significant ENDOPARASITES in reptiles.

RUGA (pl.) **RUGAE** A wrinkle, crease or fold. In a large number of *Anolis* species, a raised ridge on either side of the top of the head from the snout to the eyelid, between which lies the slightly hollowed area known as the INTERRUGAL SPACE.

RUGOSE Wrinkled or creased.

RUNNY NOSE SYNDROME A persistent discharge from the nose in terrestrial chelonians caused mostly by GRAM-NEGATIVE pathogens such as *pseudomonas, klebsiella* and *citrobacter*. Infected animals must be quarantined as certain forms of the disease are highly infectious. Topical therapy with a broad spectrum, injected antibiotic produces good results.

RUNT A dwarfed or undersized hatchling or newborn reptile; the smallest, least developed of a litter.

RUPICOLOUS Living on, or amongst, rocks.

S

SAC 1. A usually narrow-mouthed, pouch-like part in the body of a reptile or amphibian. **2.** Acronym denoting SPECIAL AREA OF CONSERVATION.

SACRAL DIAPOPHYSIS (pl.) **SACRAL DIAPOPHYSES** A transverse process on a SACRAL VERTEBRA which can, in some forms such as anurans, be expanded at its outer margin where it articulates with an ILIUM.

SACRAL HUMP *See* HUMP.

SACRAL PROMINENCE The HUMP on the lower back in anurans.

SACRAL REGION Pertaining to the area of the lower back where certain vertebrae articulate with the pelvic girdle

SACRAL VERTEBRA Any one of the strong, large vertebrae that articulate with the pelvic girdle. In reptiles there are two or more, bearing ribs which join with the ILIUM, and in amphibians a single one bearing stout processes (SACRAL DIAPOPHYSES) and ribs.

SACRUM One or more fused SACRAL VERTEBRAE attached to the ilia and providing support for the pelvic girdle. In the vertebral column of anurans, it is the last single vertebra bearing the SACRAL DIAPOPHYSES which articulate with the ILIUM.

SADDLE Descriptive of any dorsal blotch of colour in the pattern of certain snakes, that extends down the sides but is wider along the mid-line than it is laterally

SAHEL ZONE A region of SAVANNA and arid scrub lying directly south of the Sahara, approximately between 12 and 16°N.

SALAMANDER TOXIN Any of several secretions (e.g., SAMANDARIN) produced in the skin glands of urodeles, causing cramps, irritation to the mucus membranes and the skin, convulsions and paralysis.

SALAMANDRIDAE Fire salamanders, newts and related forms. Family of the URODELA, inhabiting the HOLARCTIC. Approximately 48 species in some 15 genera.

SALIENTIA The order of the Amphibia, more commonly known as the ANURA, containing the frogs and toads. With the exception of the polar regions and some oceanic islands, representatives of the order are found throughout the world.

SALIVA A watery digestive secretion produced by the salivary glands in the mouth and discharged into the buccal cavity where it performs a lubrication function, easing the passage of food into the oesophagus. The venom glands of certain snakes are modifications of salivary glands.

SALMONELLOSIS A common communicable disease occurring in reptiles, especially snakes and chelonians, caused by bacteria of the genus *Salmonella*, and often the cause of food poisoning in humans who have been careless in their hygiene practices following contact with an infected reptile. Many chelonians, especially the semi-aquatic forms, carry the *Salmonella* bacteria without showing any outward symptoms of the disease.

SALTATORIAL; SALTATORY Descriptive of limbs that are modified for leaping.

SALT GLAND Any of the glands in reptiles used for excreting concentrated excess salt solutions (OSMOREGULATION). In crocodiles, salt glands are located in the tongue (there are no such glands in alligators or caimans). In certain terrestrial lizards (e.g., *Iguana, Dipsosaurus, Ctenosaura*), the marine iguana (*Amblyrhynchus*) and the marine snakes (Hydrophiidae), they are located in the nasal passages, and in the marine turtles (Cheloniidae, Dermochelyidae), in the region of the eye.

SAMANDARIN; SAMANDARIDIN A toxic alkaloid secretion produced in the skin glands of *Salamandra* urodeles. Used as a defence mechanism, it can cause muscular cramps and convulsions in predators.

SANDFISH Any one of a group of small-legged, desert-dwelling SKINKS which hold their limbs flat against their bodies and use SERPENTINE MOVEMENT when burrowing or moving rapidly.

SAND-SWIMMING The term applied to the rapid undulatory movements of many desert-dwelling lizard species (e.g., *Angolosaurus, Anniella, Aporosaura, Uma*) as they burrow their way into sand.

SAPROLEGNIASIS A disease seen in captive aquatic amphibians and amphibian larvae, and usually a result of injury to the skin. It is similar to the disease in fish in which pale tufts of fungus appear on the skin. The most common species isolated have been *Saprolegnia ferax* and *S. parasitica,* the former being the cause of high mortality rates in spawn of Northwest American populations of the western toad *Bufo boreas.*

SARAFOTOXIN A powerfully toxic component of the venom of African mole vipers (*Atractaspis* sp.) affecting the walls of the aorta and coronary blood vessels which are narrowed, causing acute vasospasm and possible death from coronary insufficiency.

SARCOPTID MITE *See* MITE.

SAURIA The suborder of the SQUAMATA, also known as the Lacertilia, containing the lizards, the more primitive of which are arboreal or terrestrial four-legged forms whilst many of the more advanced families have evolved subterranean, limbless forms. Some 18 families are distributed widely across the world with the exception of the polar regions. Autotomy in many species is common and the epidermis is generally sloughed in pieces. Lizards became apparent and began to branch out in the TRIASSIC, and from the late CRETACEOUS onwards continental deposits have surrendered enough lizard fossils to enable the development of the group to be followed up to the present day.

SAUVAGE FINISH In the commercial skin trade, the term applied to the matt, textured, oiled-leather appearance of a specially processed crocodilian skin.

SAVANNA Or savannah. A large, open tract of grassland, with scattered trees and bushes, typical of much of tropical Africa, but also occurring in South America and Australia. Reptiles and amphibians inhabiting savannas experience an arid climate with a long dry season and short wet season.

SAXATILE; SAXICOLINE; SAXICOLOUS Living among rocks; inhabiting rocky locations.

SAY, Thomas, (1787-1834), American zoologist and co-founder of the Philadelphia Academy of Natural Sciences. Described many new North American reptile and amphibian species.

SCALATION Pertaining to the number, shape, pattern, arrangement and location of the scales in reptiles.

SCALE Any one of the flattened, horny, thick or thin, epidermal plates covering the bodies of reptiles. *See* LAMINA; PLATE; SCUTE; SHIELD.

SCALE-BOSS *See* KNOB.

SCALE-CLIPPING A method of marking reptiles, especially snakes, by clipping or cutting one or more scales (usually on the VENTER) in a predetermined combination, in order that individuals can be recognised both in the field and in the laboratory. *See* TOE-CLIPPING.

SCALE ERECTION The ability, in certain lizard species to lift the edges and/or tips of their scales away from the surface of their body in order to defend themselves against predators or to make themselves less palatable. The habit may also enable them to wedge themselves into crevices thereby rendering their extraction impossible.

SCALE FOSSA *See* SCALE PIT.

SCALE ORGAN In lizards, minute structures on the upper surface of the scales consisting of tiny rounded projections and/or hair-like filaments, apparently tactile in function.

SCALE PIT; APICAL PIT; SCALE FOSSA The tiny depressions on the apex of the dorsal scales in snakes and lizards.

SCALE ROT A form of DERMATITIS in captive reptiles, especially snakes, usually a result of keeping the animals too wet, or the vivarium too humid. In its early stages, it may appear as small blisters and/or a pinkish or brownish discoloration of the ventral scales. Untreated it very often becomes NECROTIC, i.e., areas of tissue die and slough away.

SCALE ROW The DORSAL scales of most snakes are arranged in a continuous series of straight, distinct, longitudinal and oblique rows, the number of which usually varies from one end of the body to the other. When undertaking a scale-row count, the rows at mid-body are those commonly considered. In the longitudinal type of scale row, common to most snakes, the rows run in three directions, one diagonally up from the ventrals and forward to the centre of the back, one diagonally up from the ventrals and backward to the centre of the back, and one extending the length of the body. The number obtained varies from one species to another but usually remains constant within a species. Most scale-row counts result in odd numbers, e.g., 17-19, which signifies either 17 or 19 and not 17, 18 or 19.

SCALE-ROW COUNT *See* SCALE ROW.

SCANSOR A modified pad on the tip of the toes of many geckos, made up of numerous microscopic hair-like structures, which permit the lizard to climb apparently smooth vertical surfaces by taking advantage of minute flaws and cracks within those surfaces.

SCANSORIAL Specially adapted for climbing as, for example, lizards of the family CHAMAELEONIDAE, and many geckos.

SCAP To capture freshwater or marine chelonians with the aid of a dip net.

SCAPULA (pl.) **SCAPULAE** The dorsal part of the shoulder girdle.

SCAPULAR CHEVRON A dark V-shaped bar or band transversing the dorsum above the forelimb insertion, i.e., over the site of the scapulae.

SCAPULAR REGION The area on the dorsum of a reptile or amphibian, lying directly behind the neck and over the position of the dorsal part of the pectoral girdle, or scapulae.

SCAT; BOLUS A small pellet of faecal matter.

SCENT GLAND The ANAL GLAND situated in the base of the tail of snakes, or any corresponding gland in crocodilians, chelonians, or amphibians. *See* CLOACAL GLAND; MUSK GLAND.

SCENTING *See* SCENT MANIPULATION.

SCENT MANIPULATION A method used to encourage a captive reptile to eat a food item which is not its preferred or natural food, by applying the smell of one type of food to another. For example, toads are the principal food of the North American hog-nosed snakes (*Heterodon*), but in captivity they will take rodents which have been rubbed with a toad

SCENT TRANSFERENCE *See* SCENT MANIPULATION.

SCHMIDT, Karl Patterson, (1890-1957), American biologist and herpetologist at the American Museum of Natural History in New York. In 1922 he established the Herpetological Department in Chicago's Field Museum of Natural History. Contributed many publications on herpetology including *The American alligator* (1922), *Crocodiles* (1944), *Crocodile hunting in Central America* (1952), *The truth about snake* stories (1955), and *Living reptiles of the world* (1957) which he co-authored with Robert Inger.

SCIENTIFIC NAME The title applied to each taxonomic division, understood throughout the world and serving to avoid the complications associated with the usage of many different vernacular names by providing an internationally agreed uniform system. The application of scientific names to organisms is termed NOMENCLATURE.

SCINCIDAE Skinks. Family of the SQUAMATA, suborder SAURIA, inhabiting the tropics and warmer temperate zones of the world, with the greatest number of species occurring in Australasia, South-East Asia and Africa. Estimated at around 1000 species in some 50 genera.

SKINK Any lizard of the family SCINCIDAE.

SCOLECOPHIDIA One of two major lineages to which snakes belong. containing some 300 species. Scolecophidians are all well-adapted specialized burrowers and there is much debate as to whether they are a more primitive group than the HENOPHIDIA or just superbly adapted to their burrowing way of life. *See also* ALETHINOPHIDIA.

SCOLECOPHIDIAN *See* SCOLECOPHIDIA.

SCOTOPHASE The phase of darkness during a 24 hour period of light and dark. *See* PHOTOPERIOD.

SCOTOPIA The ability of the eye to adjust for night vision.

SCREAM The distress call, or FRIGHT CRY, of an anuran.

SCREE An area of weathered rock fragments at the foot, or on the slopes, of a mountain or cliff, offering an ideal habitat for several SAXATILE species of reptile.

SCRUB An area of arid land covered with bushes, stunted trees and low vegetation.

SCUTE Any one of the enlarged scales on a reptile and alternatively termed a SHIELD or PLATE.

SCUTELLATION Pertaining to the number, shape, pattern, arrangement and location of the LAMINAE on the CARAPACE and PLASTRON of chelonians.

SCUTELLUM (pl.) **SCUTELLA** Any small scale on the body of a reptile.

SEAM The joined edges of, or border between, adjacent plates or laminae.

SEARCHING The actions of a reptile or amphibian that are focussed on the seeking of an essential item (e.g., food, or a mate) that has yet to be found.

SEASONAL BREEDER A reptile or amphibian species that, in captivity, will only breed during a particular time of the year, as opposed to other species that may breed at any time.

SEASONAL MOVEMENT The movement, or migration, of a reptile or amphibian population from one area to another, closely associated with the cycle of the seasons and usually in response to increasing or decreasing day lengths (*See* PHOTOPERIOD). The purpose behind such seasonal movement in reptiles and amphibians is usually linked to either reproduction or hibernation and, in the case of many amphibians and the marine turtles, can often involve long distances as they travel to their breeding sites. In many temperate reptile species seasonal movement occurs at the onset of winter as they migrate to an established over wintering quarter, or HIBERNACULUM.

SECONDARY CONSUMER A CARNIVORE (e.g., a python) that preys upon a HERBIVORE (e.g., a rodent).

SEDENTARY Motionless; inactive.

SEGMENT; BELL; BUTTON Any one of the parts into which something is separated, e.g., an individual keratinized ring that interlocks with other rings on the tail of a rattlesnake to form the RATTLE.

SELVA Dense tropical RAINFOREST, especially in the region of the Amazon, characterized by very tall, broadleaved evergreen trees, lianas, epiphytes and vines, and experiencing heavy rainfall most of the year and a dry season lasting only two or three months.

SEMATIC Descriptive of the conspicuous coloration typical of many poisonous amphibians and certain venomous reptiles, serving as a warning to other animals.

SEMI-AQUATIC Descriptive of any reptile or amphibian that occurs in a watery environment, or spends part of its life in water and part on land.

SEMI-ARBOREAL Descriptive of any reptile or amphibian that spends part of its life in trees.

SEMI-ARID Descriptive of any region characterized by sparse, scrubby vegetation and very limited rainfall.

SEMINAL PLUG; COPULATORY PLUG 1. Dried semen that adheres to the invaginated hemipenes of a male snake, verifying that the reptile is reproductively active, i.e., producing sperm. **2.** Dried semen that obstructs the female's genitalia, thwarting other males from depositing challenging sperm. The bigger the male's HEMIPENIS the more successful the plug.

SEMINAL RECEPTACLE *See* SPERMATHECA.

SEMI-STERILE MAINTENANCE Referring to the husbandry of captive reptiles and/or amphibians when housed in vivaria with only essential furnishings, e.g., hide box, water trough and a very simple substrate such as newspaper or bark chippings, allowing the maintenance of optimum hygiene conditions.

SEMI-TERRESTRIAL Descriptive of any reptile or amphibian that spends only part of its life on land or at ground level.

SEPTAL RIDGE A lateral ridge on the SEPTUM, or dividing wall, of the nostrils, as in the soft-shelled turtles (*Trionyx*).

SEPTICAEMIA Blood poisoning. Symptoms in humans include weakness, vomiting, collapse, unconsciousness and sometimes delerium. Generalized septicaemia can occur in reptiles as a secondary result of any bacterial infection but is particularly common in connection with necrotic DERMATITIS and ABSCESS. In anurans bacterial septicaemia is responsible for high mortality rates and is often the result of infection by *Aeromonas* *hydrophila* and other gram negative bacteria such as *Pseudomonas* and *Proteus* spp. for example.

SEPTUM A dividing partition separating two tissue masses or cavities, e.g., the vertical tissue separating the nasal passages.

SERINE An amino acid present in many proteins.

SEROTHERAPY The application of a serum, such as an antivenin when treating the bite of a venomous snake, for example.

SEROTINAL Pertaining to the late summer. *Compare* AESTIVAL; AUTUMNAL; HIBERNAL; PREVERNAL; VERNAL.

SERPENT A literary or vernacular term for a snake.

SERPENTARIUM Any building housing a collection of snakes, but often other reptiles and sometimes amphibians also. It may be a simple public display of various species or a more scientific establishment in which specimens are kept and bred for specific purposes, e.g., for venom extraction in the production of serum.

SERPENTES The suborder of the SQUAMATA, containing the elongated, limbless, snakes which evolved from subterranean burrowing lizards during the CRETACEOUS. The suborder, which is divided into approximately 12 families, is also known as the Ophidia. Snakes are carnivorous and their jaws which are secured by ligaments are capable of substantial expansion to permit the swallowing of large prey items. They lack eyelids, the eyes instead being covered by a transparent scale, or SPECTACLE, and the scent-receptive tongue is usually forked. The epidermis is normally sloughed in one piece and the tail cannot be regenerated if lost.

SERPENTINE MOVEMENT; UNDULATING MOTION The most common method of progress in snakes, in which the reptile proceeds forwards by means of continuous, rhythmic, winding, lateral waves of the body. A method of movement also employed by other legless forms, e.g., certain lizards and caecilians.

SERUM *See* ANTISERUM.

SETA (pl.) **SETAE** Any hair or hair-like appendage; a fine, hair-like structure located on the under surface of the feet of certain geckos, which enables them to adhere to smooth vertical surfaces. Each gecko foot may have up to half a million setae, with each seta being less than a tenth of the thickness of a human hair. At the tip of each of these setae are situated yet smaller mushroom-shaped filaments called spatulae, ten million of which could fit on a pin head. Consequently, each gecko foot has a billion of these miniscule filaments.

SEX CALL A form of VOCALIZATION in male anurans, which acts as a mating invitation to females, and on to which only members of the same species will home in.

SEX DETERMINATION The ascertainment of sex in a reptile or amphibian. The methods used include superficial examination to detect differences in the size, shape or colour of individuals, or the presence of any one or more structures or appendages associated with male or female sex characteristics, such as crests, vocal sacs, preanal and/or femoral pores, or dewlaps, for example. If no such features are present then it may be necessary, in the case of snakes in particular, to use a CLOACAL PROBE to determine an individual's sexual identity.

SEX RATIO The number of males in relation to females as they occur in the wild or in captivity. For example, male leopard geckos *Eublepharis macularius* are very territorial and may inflict injuries upon each other if housed together and are, therefore, usually kept in a sex ratio of one male to two or more females.

SEX SEPARATION A frequently used method of encouraging reptiles, especially snakes, to breed by first isolating the males from the females for a short period of a few weeks, thereby overcoming the sometimes non-productive 'familiarity' that can occur between animals constantly housed together.

SEXUAL DIMORPHISM The morphological condition in which the male and female of a species exhibits distinct differences in colour, markings, size and/or structure. Sexual dimorphism is evident, in one form or another, in many reptiles, especially lizards and chelonians, and in some amphibians.

SHEATH A membrane enfolding or enveloping a structure. The fang sheath surrounding the base of a fang helps direct venom into the fang's venom canal.

SHED; SLOUGH To cast off, or moult, the outermost layer (STRATUM CORNEUM) of skin. *See* ECDYSIS; EXUVIATION.

SHELFORD'S LAW OF TOLERANCE A law asserting that the abundance or dispersal of a creature can be controlled by specific features (e.g., climate, topography, and biological factors) where levels of these go beyond the maximum or minimum LIMITS OF TOLERANCE of that creature.

SHELL In the majority of chelonians, the hard, rigid structure enclosing the entire body with the exception of the head, limbs, and tail, although in most species these are retractile to a varying degree, consisting of bony plates overlaid with enlarged scales, or laminae. In a few chelonians (TRIONYCHIDAE), the structure is much softer and covered with skin. The shell consists basically of two parts, the CARAPACE and the PLASTRON, united on either side of the body between the front and hind limbs by the BRIDGE.

SHELL-BREAKER The horny outgrowth, or CARUNCLE, on the tip of the snout in baby chelonians, tuataras and crocodilians, for slitting the shell of the egg prior to hatching. *Contrast* EGG TOOTH.

SHELL CALCIFICATION In OVIPAROUS reptiles, eggs are shelled several days before deposition as they pass through the OVIDUCTS on their way to the coelonic cavity. The amount of calcification varies between species and can be affected by METABOLIC BONE DISORDER resulting from dietary or environmental deficiencies when GRAVID females have not received supplemental calcium in their diet.

SHIELD In reference to chelonians, any one of the large horny plates (LAMINAE) that cover the SHELL or, as used by some authors, the shell itself. Sometimes also used for any one of the enlarged scales, or PLATES, on the heads of many snakes.

SHOCK A disturbance of the oxygen supply to body tissues and return of blood to the heart, often a result of HYPOTENSION. *See* ANAPHYLAXIS.

SHOVEL The SPADE on the hind foot of the spadefoot toads (*Scaphiopus, Pelobates*).

SIB The abbreviated form of SIBLING.

SIBLING The progeny of the same parents, but not necessarily of the same litter.

SIDE HIDE In the commercial skin trade, the term applied to the narrow piece of skin cut from beneath the lower jaw, running backward to transverse the foreleg, extending along the flank, beneath the hind leg, and stopping close to the vent. Such strips are usually taken from the larger adult caimans (*Paleosuchus, Melanosuchus, Caiman*) all of which have belly hides that are of no use as leather due to the presence of heavy osteoderm buttons.

SIDE ORGAN Any organ on the side of the body of aquatic amphibians, forming the LATERAL LINE SYSTEM.

SIDEWINDER Any desert-dwelling snake which uses SIDEWINDING MOVEMENT in order to travel across loose sand. *Crotalus cerastes*, a snake of the VIPERIDAE family, is one example, inhabiting the sandy arid areas, desert sand dunes, and sandy hillsides, especially in areas of creosote or mesquite shrubs, of western and central Sonora, Mexico; northward to southern central Arizona

SIDEWINDING MOVEMENT; CROTALINE MOVEMENT The method of progress used by certain desert-dwelling snake species (e.g., *Cerastes vipera, Crotalus cerastes*) when traversing loose sand, by means of sideways looping movements of the body.

SILURIAN The geological period of the PALAEOZOIC era, beginning some 440 million years ago at the end of the ORDOVICIAN and extending for about 35 million years, when it was succeeded by the DEVONIAN period.

SINCIPITAL The collective name for any FRONTAL or PARIETAL scale on the heads of reptiles.

SINCIPUT The anterior upper part of the head, or skull. *Contrast* OCCIPUT.

SINUATE Curved; having an undulating or indented border.

SINUS Any hollow depression or

cavity in the body.

SIPHONIUM A narrow, membranous tube in the head of crocodilians linking the cavity of the middle ear with the articular bone of the lower jaw.

SIREN *See* SIRENIDAE.

SIRENIDAE Sirens. Family of the URODELA, inhabiting the south-eastern NEARCTIC. Three species in two genera. Often placed within its own order, the Trachystomata.

SITE OF SPECIAL SCIENTIFIC INTEREST; SSSI An area, particularly in the British Isles, which is of particular importance and worthy of protection, especially from damaging operations, due to its floral, faunal, geological, or physiographical features. An SSSI is not necessarily owned by the Government or a conservation organization but can indeed be owned by anybody.

SKIN BLISTER *See* BLEB; BULLA; DERMATITIS; THERMAL BURN.

SKIN FRINGE Any fleshy or scaly margin on the body of a reptile, especially on the posterior borders of the limbs.

SKINK Any lizard of the family SCINCIDAE most common in tropical Africa and Asia. Characteristic features include reduced limbs and an elongated body covered with smooth scales.

SKITTER To move rapidly or lightly, especially over water; used to describe the flight of certain anuran and lizard species across a body of water when startled or pursued.

SLAP-FEEDING A technique used to encourage snakes which are reluctant feeders to strike and seize a food item which is tapped lightly against the snake's snout.

SLIDE-PUSHING MOTION An alternative term for CONCERTINA MOVEMENT.

SLIDER *See* COOTER.

SLOUGH The moulted or cast-off outermost skin layer of a reptile or amphibian. *See* ECDYSIS; EXUVIATION; SHED.

SLOUGH CYCLE; ECDYSIS CYCLE The period in a reptile's life between one skin SLOUGH and the next, from the time it has successfully completed a shed to when it becomes MILKED-UP prior to the next slough.

SLUG Any unfertilised ovum passed by a female reptile and frequently pushed away from the main clutch. Slugs are generally yellowish in colour and are merely yolk and shell.

SMALL-SCALE HIDE In the commercial skin trade, the term applied to soft-bellied crocodilian species possessing 26-35 TRANSVERSE VENTRAL SCALE ROWS. *See* LARGE-SCALE HIDE.

SMCC Acronym denoting the Species Management Coordinating Council (ARAZPA and COGBAZ).

SMITH, Malcolm Arthur,(1875-1958), British physician and herpetologist, and research associate at the British Museum (Natural History), London. Played a significant role in the development of herpetology in Britain and helped to establish the British Herpetological

Society. Over 120 publications including *Monograph of the seasnakes* (1926), *The breeding habits of the Indian cobra* (1937), and the classic *The British amphibians and reptiles* (1951).

SMCC Acronym denoting the Species Management Coordinating Council (ARAZPA and COGBAZ).

SMP Acronym denoting the Species Management Programme (Australian version of SSP).

SNAKE ANTITOXIN *See* ANTISERUM.

SNAKE BITE Pertaining to any bite of a snake, but especially that of a venomous species, or the condition arising from the bite of a venomous snake.

SNAKE BITE SERUM; ANTISERUM; ANTIVENENE; ANTIVENIN; ANTIVENOM A serum, processed from the blood of an animal that has received gradually increasing, immunizing doses of a VENOM and/or ANAVENOM, which has the ability to counteract or neutralize the effects of a specific snake venom.

SNAKE CHARMER Any entertainer, especially in parts of Asia (*See* SNAKE WALLAH), who, with swaying movements of the body and by playing music, encourages certain snakes, particularly cobra, in his possession to respond in certain ways, thereby giving the impression that they are being magically charmed.

SNAKE FARM Any establishment, usually in a tropical or subtropical country, set up to keep and breed snakes, especially venomous species. *See* SERPENTARIUM.

SNAKE HOOK An instrument, varying in size and design but generally consisting of a long pole serving as a handle with an L-shaped piece at one end, suitable for pinning snakes to the ground or lifting dangerous species whilst keeping them at arms length.

SNAKE ISLAND The name given to Ilha de Queimada Grande, a tiny island off the Brazilian south-east coast on which is rumoured to be dangerous snakes, the population of which is so dense that there are five to every square metre. The particular snake in question is the golden lancehead *Bothrops insularis*, a unique species of FER DE LANCE pit viper with a tremendously powerful venom. Despite the rumours, and because of its island isolation, it is now considered to be at great risk of extinction.

SNAKE MITE *Ophionyssus natricis*, a minute arachnid of the order Acarina, frequently infesting snakes. They may occur on any part of the snake's body but tend to concentrate around the head, especially the eyes and nostrils. Snake mites suck the blood of their host and heavy infestations can often result in a snake's death.

SNAKE ROOT Any one of several North American plants (e.g., *Eupatorium urticaefolium, Aristolochia serpentaria*), the rootstalks of which have been used to cure SNAKE BITE.

SNAKE STONE Any porous object

which, when placed on the punctures of a snake bite, is supposed to draw out and absorb the venom. The BEZOAR STONE found in the stomachs of certain mammals is an example.

SNAKE WALLAH A term, used in India and Pakistan, for a person who cares for, or is in charge of, snakes; a snake-charmer.

SNARING *See* NOOSING.

SNIFFLES In reptiles, especially snakes, a discharge from the nasal passages, resulting from inflammation of the mucous membranes, and usually the result of chilling. The watery discharge may also be accompanied by expiratory wheezing noises and nasal bubbling.

SNORKEL The lengthened tubular snout, or PROBOSCIS, of certain aquatic chelonians, which enables them to breathe whilst the rest of their body is completely submerged.

SNOUT The part of the head anterior to the eyes, consisting of the nostrils, jaws and surrounding area.

SNOUT-VENT LENGTH; SVL; HEAD-BODY LENGTH The straight-line length of an anuran, caecilian, snake, lizard or crocodilian, as measured from the anterior tip of the snout to the posterior margin of the anus or vent. *Contrast* TOTAL LENGTH.

SNOW A fairly recently devised term for the overall whitish corn snake *Pantherophis guttatus* resulting from a combination of the AMELANISTIC and ANERYTHRISTIC genetic mutations.

SOFT-SHELL A deformity in the CARAPACE of a chelonian, as a result of a calcium deficiency (HYPOCALCEMIA) which, in captive animals, is usually linked to incorrect nutrition and/or a calcium: phosphorus imbalance in the diet.

SOLENOGLYPH A snake that possesses venom-conducting fangs connected to a comparatively kinetic MAXILLA bone, enabling them to be folded back against the roof of the mouth. *Contrast* PROTEROGLYPH.

SOLENOGLYPHIC TEETH The venom-conducting teeth of front-fanged vipers, hollow and situated well forward on the upper jaw.

SOLE TUBERCLE In many anurans, a METATARSAL TUBERCLE on the hind foot.

SOMATOLYSIS Disruptive pattern and coloration exhibited by many reptiles and amphibians which, in the animal's normal environment, breaks up its body contours making it difficult if not impossible to see but, when seen against a plain background, renders the animal extremely conspicuous. The gaboon viper *Bitis gabonica* is an excellent example of somatolysis.

SOOGLOSSIDAE Seychelle frogs. Family of the ANURA, inhabiting the Seychelle Islands. Three species in two genera.

sp. Abbreviation for SPECIES (sing.).

SPADE The enlarged METATARSAL TUBERCLE on the hind foot of spadefoot toads (*Scaphiopus, Pelobates*), used for digging. Also, but less commonly, termed the 'shovel'.

SPARGANA The sub-adult stage of CESTODES.

SPARKS Acronym denoting the Single Population Analysis and Record Keeping System. SPARKS is computer software employed by studbook keepers and population managers to carry out the essential genetic and demographic analysis required in the management of species held in captivity.

SPASM A sudden involuntary muscular contraction, especially one that results in convulsion or cramp.

SPATULA (pl.) **SPATULAE** *See* SETA.

SPATULATE Flat and rounded at the tip; shaped like a spatula, as for example, the large, flattened toe DISC of many arboreal anurans.

SPAWN The gelatinous mass or string of eggs produced by amphibians. *See* EGG MASS.

SPECIAL AREA OF CONSERVATION; SAC The European Union's *Species and Habitats Directive* has identified some 169 habitat types and over 620 species (listed in its Annexes I and II) which they aim to conserve by establishing a succession of high quality SACs throughout Europe.

SPECIALIST An animal having a diet consisting of one or more specific foods only.

SPECIALIZATION The level of modification of a creature to its surroundings. A high degree of specialization advocates both a limited habitat or role and important interspecific competition

SPECIATION The development of one or more new species from an existing species, as when a geographically isolated population evolves distinctive features as a result of natural selection, and can then no longer interbreed with the parent population.

SPECIES A group of similar individuals that are able to breed among themselves and produce offspring. Individuals or groups belonging to the same species are termed 'CONSPECIFIC'. Each species has a SCIENTIFIC NAME designated in italics by the GENUS name followed by the specific name, e.g., *Python regius* (the royal python). Similar or related species are placed within a genus; many species are divided into SUBSPECIES. *See* BIOLOGICAL SPECIES; EVOLUTIONARY SPECIES; TAXONOMIC SPECIES.

SPECIES COMPLEX Any group of closely related species having a common and normally recent ancestor.

SPECIES DIVERSITY The different types of species occurring in a particular region or habitat. Species diversity among reptiles and amphibians is greatest within the tropics. *See* BIODIVERSITY.

SPECIES ISOLATION The separation of, or 'barriers' between, different species, serving to prevent members of one species attempting to interbreed with members of another different species, which would result in either infertile mating or sterile offspring, both of which would threaten the continued existence of both species. Species isolation is achieved through various mechanisms including olfactory signals (via glandular secretions), visual signals (species-specific behaviour) and acoustic signals, as in the pre-mating calls of anurans. *See* PRE-MATING ISOLATING MECHANISMS.

SPECIES-SPECIFIC Restricted to one particular species.

SPECIFIC NAME In the SCIENTIFIC NAME of a given reptile or amphibian, the second word in italics, indicating the SPECIES. For example, in the name *Spalerosophis diadema* (the diadem snake), the specific name is *diadema*.

SPECTACLE *See* BRILLE.

SPERMATHECA; RECEPTACULUM SEMINIS; SEMINAL RECEPTACLE The receptacle or cavity in the cloaca of female urodeles, that receives the male's SPERMATOPHORE and holds it until fertilization occurs.

SPERMATOGENESIS The formation and development of sperm within the male testes.

SPERMATOPHORE The gelatinous packet or capsule containing sperm, secreted by cloacal glands and deposited by a male urodele in water or on moist soil, where it is then taken up by a female and retained in a cavity (SPERMATHECA) within her cloaca until fertilization takes place.

SPERMATOZOON (pl.) **SPERMATOZOA** Sperm; the mature, motile reproductive cell (GAMETE) produced by the male

testes.

SPERM CAPSULE The gelatinous packet (SPERMATOPHORE) containing the sperm of a male urodele.

SPHENODONTIDAE Tuataras. Only family of the RHYNCHOCEPHALIA, suborder Sphenodontida, inhabiting a few islands off New Zealand. MONOTYPIC genus.

SPHENOIDAL TEETH In skinks of the genus *Eumeces*, teeth situated on the PTERYGOID bone of the skull.

SPIDER-WEB UMBILICUS A term used for the network of lines and creases in the soft, scale-less skin surrounding the rear part of the umbilicus scar on the venter of the American alligator (*Alligator mississipiensis*).

SPINAL Of, or relating to, the spine or spinal cord. Sometimes used in place of VERTEBRAL when describing scales or markings along the mid-dorsal line of a reptile.

SPINAL CORD The part of the central nervous system of vertebrates, connecting the brain and the nerve cells that supply the muscles and organs of the body by means of a series of paired spinal nerves along its length. In reptiles and amphibians it extends from the brain to the tail, enclosed within the neural canal of the vertebral column.

SPINE Any firm, pointed structure or process on the body of a reptile (e.g., above the eye in the African horned viper, *Cerastes cerastes*, and on the tail tip of uropeltid snakes) or amphibian (e.g., on the dorsum of the

base of the tail in the spine-tailed salamanders, *Mertensiella*, and the skinless innermost digit on the forefoot of some anurans).
See TAIL SPINE.

SPINOSE Bearing numerous spines.

SPIRACLE; BRANCHIAL APERTURE; BREATHING PORE The small external opening in the body wall of an anuran tadpole that leads to and from the gill chambers (atria). In most anuran species there is a single spiracle situated on the left side of the body of the tadpole. In the LEIOPELMATIDAE, DISCOGLOSSIDAE and MICROHYLIDAE it is placed mid-ventrally and, in the RHINOPHRYNIDAE and PIPIDAE, it is paired bilaterally.

SPITTER Herpetological jargon for any one of several elapid snake species, e.g., the ringhals *Hemachatus haemachatus* and the black-necked cobra *Naja nigricollis*, that possesses the ability to expel venom from the mouth in two fine streams, one from each fang, at an aggressor.

SPLINTER SCAR Evidence of a wound on the CARAPACE of a chelonian, in which a long thin strip of the bony layer has been exposed between the surrounding keratinous laminae.

SPONTANEOUS FEEDING The natural, voluntary feeding (i.e., unassisted or unforced) in a reptile, especially a hatchling or newborn individual. *Contrast* FORCE-FEEDING.

SPOROZOA Parasitic unicellular organisms which reproduce by forming spores, and commonly responsible for blood and intestinal infections in reptiles, and for small tumours and cysts in the skin and muscles of amphibians.

spp. Abbreviation for SPECIES (pl.).

SPREADING The flattening of the neck to its fullest extent, characteristic of an angry or excited cobra, but also seen to a slightly lesser degree in a few other species, e.g., the 'false water cobras' (*Hydrodynastes*) and the hognosed snakes (*Heterodon*).

SPUR Any sharply pointed, rigid, spine-like structure often on the hind limbs, as in the spur-thighed tortoise (*Testudo graeca*) and certain urodeles (*Euproctus*), or on either side of the vent, as in the vestigial hind limbs of certain primitive snake species.
See ANAL SPUR; PELVIC SPUR.

SPUR FLAP A succession of enlarged, connected scales along the outer border of the forelimb in the Australian swamp turtle *Pseudemydura umbrina*, giving the appearance of a flap.

SPUR STIMULATION The use, by a male boid snake, of the ANAL SPURS during courtship to stimulate a female into sexual activity. Scratches on the female's body in the region of the cloaca may often be a sign that she has been engaging in mating activity.

SQUAMA A scale, or scale-like structure.

SQUAMATA Scaled reptiles. Accounting for 95% of all living reptiles, the Squamata is the largest order of the Reptilia, containing the most successful living reptiles, the lizards (SAURIA, or Lacertilia) and the snakes (SERPENTES, or Ophidia), and characterized by an external covering of small, horny, overlapping scales and a forked or notched tongue. The lizards contain many limbless forms.

SQUAMATE A member of the SQUAMATA; a snake or lizard.

SQUAMATION The state of possessing, or developing, scales and, in reptiles, the arrangement, shape and number of scales on the body.

SQUAMOSAL Either one of a pair of thin, plate-like bones on the side of the skull, overlain by the TEMPORAL scales.

SQUAMOUS Scaly; covered with, having the appearance of, or formed from, scales, e.g., the 'squamous HORN' on the snout of the African rhinocerous viper *Bitis nasicornis*.

SQUIRT GLAND A dermal gland in caecilians producing a liquid, strongly irritant to the mucous membranes, which can be forcibly expelled by muscular contraction.

ssp. Abbreviation for SUBSPECIES.

SSP Acronym denoting the Species Survival Plan. The plan was established in 1981 as a cooperative population management and conservation program for certain species at the American Zoo and Aquarium Association.

SSSI *See* SITE OF SPECIAL SCIENTIFIC INTEREST.

STABILIZER; BALANCER; HALTERE Either one of a pair of short, elongated projections or stalks situated on the mandibular arch of certain urodele larvae (e.g., Ambystomidae, Hynobiidae, Salamandridae) during the early stage of their aquatic life, serving to maintain balance whilst swimming, and disappearing with the growth of the forelimbs.

STAFF OF AESCULAPIUS An emblem, used by the Royal Medical Corps, the Royal Canadian Medical Corps, and the American Medical Association, consisting of a rod around which is entwined a serpent. *Compare* CADUCEUS (sense **2**.)

STAGHORN GILL The gill of the terrestrial larvae of certain urodeles, named after its striking similarity to the antlers of an adult male deer.

STASIS TADPOLE An anuran larva that has been deprived of food during its premetamorphic stage, resulting in a cessation of, or stagnation in, its development.

STATUS The position of a reptile or amphibian species in a given area, indicating the frequency with which it appears, and frequently described imprecisely by such adjectives as 'common', 'scarce' or 'locally abundant'.

STAUROTYPIDAE Mexican musk turtles. Family of the CHELONIA, suborder Cryptodira, inhabiting Central America from Mexico to Honduras. Three species in two genera.

STEINDACHNER, Franz, (1834-1919), Austrian zoologist at the Imperial Museum of Vienna. Published much on the subject of herpetology.

STEJNEGER, Leonard, (1851-1943), Norwegian physician and zoologist at the Smithsonian Institution, Washington DC. His many publications include *The poisonous snakes of North America* (1895), *The herpetology of Japan* (1907), and *Checklist of North American amphibians and reptiles* (1917) written jointly with Thomas Barbour.

STENO- A prefix indicating narrowness or constriction and used in ecology to signify an organism's limited tolerance of certain environmental factors.

STENOECIOUS Descriptive of any species which has narrow tolerances or is very specialized. *Compare* EURYOECIOUS.

STENOPAIC Descriptive of an eye in which the PUPIL is narrow and slit-like.

STENOPHAGOUS Feeding upon a specific type, or limited range, of food items. Stenophagy occurs in a number of reptiles, e.g., the marine iguana *Amblyrhynchus cristatus*, which feeds on kelp, and the egg-eating snake *Dasypeltis scabra*, which as its name suggests feeds on the eggs of birds.

STENOTHERMAL Able to exist only within a very narrow range of temperatures. *Compare* EURYTHERMAL.

STEPPE A vast area of usually treeless grassland in Eurasia, especially the former USSR, characterized by prolonged hot summers, often with extensive drought, and cold winters with heavy snow and long periods of frost. The reptile and amphibian fauna includes several chelonian species, anurans such as bufonid and pelobatid toads, and a number of snake and lizard species.

STERNUM An unpaired, shield or rod-shaped cartilage or bone on the ventral side of the thorax, articulating with the bones of the pelvic girdle and with most of the ribs when these are present. Commonly termed 'breastbone'.

STILET Used for the often tiny, sometimes toeless and pointed limb of certain lizards, e.g., the South African large-scaled grass lizard, *Chamaesaura macrolepis*.

STILLBIRTH In reptiles exhibiting VIVIPARITY, the birth of dead, but usually fully-developed, young.

STOMATITIS; CANKER; MOUTH ROT; NECROTIC STOMATITIS; STOMATITIS INFECTIOSA; ULCERATIVE GINGIVITIS; ULCERATIVE STOMATITIS. An infection of the oral cavity in snakes and, less commonly, lizards, usually caused by GRAM-NEGATIVE BACTERIA, as a result of a mouth injury, suboptimal captive husbandry practices, or STRESS, or a combination of these and several other health-weakening conditions.

STRADDLE AMPLEXUS A type of AMPLEXUS engaged in by certain

Malagasy ranid frogs, in which the male sits astride the shoulders of a female whilst both grasp a suspended leaf.

STRATUM CORNEUM;
CORNEAL LAYER The outermost layer of dead, flattened, horny, keratinized cells of the epidermis which provides a protective barrier between the body and the environment. The stratum corneum of reptiles and amphibians is shed frequently during their life to allow for growth (*See* ECDYSIS), and may be sloughed relatively quickly and in one piece, as in snakes and amphibians, or gradually and in numerous pieces, as in lizards, crocodilians and chelonians. A useful source of protein, it is often eaten by many amphibians and certain lizards as it is shed.

STRESS A condition that disrupts the normal biochemical and physiological patterns in the body and which, in captive reptiles and amphibians, can result from any one or combination of several factors, e.g., careless or over-handling, temperatures that are too high, housing together two or more incompatible species or individuals. Stress is often evident in specimens following transportation and in recently collected wild individuals. It is the most frequent cause of illness amongst captive animals.

STRIATE Marked with narrow bands, or lines of textures or colours, that contrast with the ground colour or texture; ridged, grooved, or striped with parallel markings, as on the scales of many reptiles.

STRIDULATION The production of a harsh rasping or creaking sound in certain snake species, e.g., the saw-scaled vipers (*Echis*) and egg-eating snakes (*Dasypeltis*), when their obliquely positioned keeled scales rub against each other.

STRIKE The lunging, usually swift movement of the head and anterior part of the body with which a snake secures its prey or defends itself by biting an aggressor.

STRIKE REFLEX *See* REFLEX FEEDING.

STRIKING COIL The characteristic position assumed by many disturbed or provoked venomous snakes (and some non-venomous species) in which the front of the body is curved backward in an S-shape prior to a STRIKE.

STRING The sound-producing appendage, or RATTLE, on the tail of a rattlesnake.

STRIPE A comparatively long narrow band of contrasting colour, or texture, running along the length of the body of a reptile.

STRIPED PHASE In certain snakes a variety having longitudinal stripes running down the length of the back instead of the normal blotched or spotted pattern. STRIPING is the result of a recessive genetic mutation.

STRIPING A pattern defect in snakes in which the normal spots or blotches are joined to form stripes. Most striping results from the accidental exposure of GRAVID females to temperatures below or, less commonly, slightly above their optimum range.

STUMP-TAIL A term commonly applied to a lizard that has recently lost its tail through AUTOTOMY, and is in the process of growing a new one.

STYLE Used for the slender, pointed, degenerate limb of certain lizards, as an alternative to STILET.

SUB- A prefix meaning located beneath; under.

SUBACRODONT Teeth, situated in the upper jaw, which are replaced several times during an animal's lifetime.

SUBANGULAR GLAND A distinct rounded bulge situated on either side of the lower surface of the jaw in species of gopher tortoise (*Gopherus*), swelling during the breeding season and exuding a strong-smelling scent which is used by the males to mark out their territories and their mates.

SUBAPICAL LAMELLA Any one of the thin single or divided plates, excluding the larger distal plates, that extend across the underside of the digits in many gecko species.

SUBARTICULAR TUBERCLE A small, rounded, granular protuberance situated between two digital PHALANGES on the underside of the fore- and hind foot in anurans.

SUBCAUDAL Any one of the scales situated on the underside, or ventral surface, of the tail from the vent to the tail tip. In most snake species the subcaudals are divided, lying in a double row. In others they may be single, or partly single and partly divided.

SUBCUTANEOUS Beneath the skin, as in a 'subcutaneous injection' when administering a drug to a reptile.

SUBDIGITAL LAMELLA Any one of the single or divided scales or plates situated on the underside, or ventral surface, of the digits in most lizard species.

SUBDIGITAL PAD *See* ADHESION SURFACE.

SUBFAMILY The taxonomic category, used in the classification of organisms, that consists of a division of a FAMILY, ranking between it and the genus. The SCIENTIFIC NAME of a subfamily ends in '-inae', e.g., Viperinae, the subfamily of the family VIPERIDAE.

SUBLABIAL Any one of the scales on the lower jaw which, in snakes, occupy the area between the LOWER LABIAL scales and the CHINSHIELD and, in lizards, make up the second scale row beneath the lip.

SUBLETHAL Referring to a measure of toxin that is less than the amount necessary to kill an animal. *See* mg/kg; LD_{50}.

SUBLINGUAL Of, related to, or situated on, the area beneath the tongue, and used by some authors for the scales on the area of the chin.

SUBLITTORAL Of, related to, or inhabiting, the shallow-water zone of a lake, or of the sea between the shore and the continental shelf. The sublittoral zone of a freshwater lake refers to the depths of between 6 and 10 metres, and that of the sea to depths of between 6 and 200 metres.

SUBMARGINAL Pertaining to the ventral surface of a MARGINAL scale or LAMINA.

SUBMAXILLARY An alternative term for the CHINSHIELD of snakes and lizards.

SUBMENTAL Used occasionally for the CHINSHIELD of snakes and lizards.

SUBMENTAL GROOVE Used occasionally for the medial furrow on the under surface of the lower jaw in snakes; the MENTAL GROOVE.

SUBOCULAR Pertaining to a scale or scales located directly beneath the eye and separating the eye from the SUPRALABIAL scales. Present in the death adders *Acanthophis* spp.

SUBOPTIMAL Less than favourable or below standard. Frequently used in reference to hygiene, captive husbandry and conditions etc.; for example, 'suboptimal temperatures', meaning below the temperatures required.

SUBORBITAL BAR An elevated crease or ridge of skin beneath the eye usually extending backward to the angle of the jaw.

SUBORDER The taxonomic category, used in the classification of organisms, that consists of a division of an ORDER, ranking between it and the family, e.g., the Cryptodira is a suborder of the CHELONIA.

SUBPLEURODONT A reptile possessing laterally-situated teeth on the inner margin of the jaw which, when lost, undergo INTERCALARY REPLACEMENT.

SUBSPECIES The taxonomic category, used in the classification of organisms, that consists of a division of, and ranks below, a species. Often termed a 'RACE', a subspecies is given a three-part scientific name (TRINOMIAL NAME) which incorporates its genus, species and subspecies.

SUBSPECIFIC NAME *See* TRINOMIAL NAME; SUBSPECIES.

SUBSTRATE The surface upon or in which a reptile or amphibian lives, either in its natural environment, or in captivity.

SUBTERRANEAN; SUBTERRESTRIAL Living and functioning below ground. Although many reptile and amphibian species may spend part of their lives below ground during HIBERNATION or AESTIVATION, some are truly subterranean and are rarely seen above the surface. Examples include members of the CAECILIIDAE, TYPHLOPIDAE and UROPELTIDAE.

SUBTERRESTRIAL *See* SUBTERRANEAN.

SUCKER; SUCKING DISC An alternative term for the ADHESIVE ORGAN around the mouthparts of embryonic anuran larvae.

SUCKING DISC *See* ADHESIVE ORGAN; SUCKER.

SULCATE Marked with longitudinal parallel grooves. Descriptive of the grooved side of a HEMIPENIS along which the sperm is conveyed. *See* SULCUS SPERMATICUS.

SULCUS A furrow, or faintly depressed linear groove.

SULCUS MARGINALIS In anurans, the furrow extending along the inner margin of the upper lip, in which the rim of the lower jaw sits when the mouth is shut.

SULCUS SPERMATICUS A distinct groove, or SULCUS, which extends along the length, and on the surface, of the HEMIPENIS in some male reptiles, along which sperm is conveyed during copulation.

SULPHONAMIDE One of a group of organic bacteriostatic compounds that prevents bacteria from reproducing and often used, with veterinary guidance, in the treatment of a variety of bacterial infections in captive reptiles and amphibians.

SUNGAZER The giant zonure or girdle-tailed lizard *Cordylus giganteus* of southern Africa, named 'sungazer' after the posture it adopts whilst basking.

SUPERFAMILY The taxonomic category, used in the CLASSIFICATION of organisms, that is part of a division of a FAMILY ranking between it and the infraorder. The SCIENTIFIC NAME of a superfamily ends in -oidea, e.g., Colubroidea, the superfamily of the family VIPERIDAE.

SUPPLEMENT DUSTING The sprinkling of a vitamin and/or mineral supplement on the food of captive reptiles and amphibians in order to boost their nutritional requirements.

SUPRA- A prefix meaning over or above; beyond.

SUPRA-ANAL KEEL In the males of some coral snake species (e.g., the Nayarit coral snake, *Micrurus proximans*), a strongly defined keel on the lateral scales of the posterior part of the body in the region of the vent, or anus. In other species (e.g., Stuart's coral snake, *M. stuarti*) they are indistinct, and in others are absent altogether. Supra-anal keels do occur occasionally on very large adult females in certain species.

SUPRACAUDAL; ANAL; CAUDAL; POSTCENTRAL; POSTMARGINAL In chelonians, the single or paired posterior laminae on the edge of the CARAPACE.

SUPRACEPHALIC Of, or related to, or situated on, the top of the head, e.g., the FRONTAL or PARIETAL scales of a snake.

SUPRACILIARY Any one of a number of small scales situated on the outermost margin of the SUPRAOCULAR scale above the orbit of the eye.

SUPRALABIAL An enlarged scale situated on the edge of the lip of the upper jaw.

SUPRALABIAL GLAND In most species of frog of the genus *Rana*, a well-defined secretory area on the upper lip extending back from below the nostril to a point posterior to the angle of the jaw.

SUPRALITTORAL Of, or related to, or inhabiting, the shore of a lake or sea above the high-water mark.

SUPRAMARGINAL In some chelonian species, any one of the laminae situated between the mid-lateral or COSTAL laminae and the MARGINAL laminae of the carapace.

SUPRANASAL Any scale or scales situated directly above the NASAL scale in snakes and lizards.

SUPRAOCULAR Any scale or scales situated above the eye and referring in particular to, in lizards, any one of the scales situated on the back of the eye socket (ORBIT) or, in snakes, the often enlarged scale directly above the eye which may, in certain species, project outwards slightly to form a SUPRAOCULAR RIDGE.

SUPRAOCULAR RIDGE; SUPRAORBITAL RIDGE The prominent elongated margin of the SUPRAOCULAR scale which overshadows the eyeball in a number of snake species, especially vipers.

SUPRAORBITAL SEMICIRCLE Used mainly in reference to a curved row of small scales situated between the SUPRAOCULAR scales and the other centrally positioned head scales.

SUPRAPYGAL; EPIPYGAL; POSTNEURAL; PROCAUDAL The term applied to either of the second and third hindmost bony NEURAL plates directly above the PYGAL bone on the mid-line of the chelonian carapace.

SUPRATYMPANIC Pertaining to the area directly above the TYMPANUM, or eardrum.

SUPRATYMPANIC FOLD; SUPRATYMPANIC RIDGE In many anurans, a fold or ridge of skin directly above, and overlying, the TYMPANUM, sometimes extending posteriorly to cover the shoulder. Especially well-developed in the Australian White's tree frog *Litoria caerulea*

SUPRATYMPANIC RIDGE *See* SUPRATYMPANIC FOLD.

SUTURE The seam or crease formed from the joining together of two surfaces, and used in herpetology for the grooved boundary occurring between adjacent laminae, plates or scales.

SVL Abbreviation for SNOUT-VENT LENGTH.

SWEEPSTAKE DISPERSAL ROUTE An expression created in 1940 by G. G. Simpson to illustrate a potential route of faunal exchange which is unlikely to be used by the majority of animals but which will, in all probability, be used by some. It necessitates a main barrier that is sporadically traversed. Which groups cross the obstruction and when they cross it are decided effectively by chance. *See* CORRIDOR and FILTER DISPERSAL ROUTES, and SEASONAL MOVEMENT.

SYMBIONT Any one of the individual organisms involved in a symbiotic relationship with other different organisms. *See* SYMBIOSIS.

SYMBIOSIS In general, the close association between two or more individuals of different species that are in some way dependent upon each other, as seen in parasitism for example, but referring in particular to a relationship in which the interactions are mutually beneficial to all the individuals (SYMBIONTS) concerned. For example, in the symbiotic relationship that occurs between the Galapagos giant tortoise *Geochelone elephantopus* and some of the island's finches, the birds obtain nourishment in the form of ticks picked from the bodies of the reptiles, which assume a particular posture to enable the birds to reach otherwise hidden places, whilst the tortoises are freed of troublesome parasites and, at the same time, warned by the birds' behaviour of any approaching enemies.

SYMPATRIC Said of closely related species that coexist within the same, or overlapping, area or region, but which do not interbreed due to one or more factors, e.g., having different breeding seasons, behavioural differences and various SPECIES ISOLATION mechanisms. *Compare* ALLOPATRIC; PARAPATRIC.

SYMPHYSIS A type of joint articulating by means of smooth layers of cartilage and fibrous ligaments, allowing only minimal movement, e.g., the joint formed between the two halves of the lower jaw of a chelonian.

SYMPTOMATIC Of, being, relating to, or suggestive of a symptom

SYNAPSE The point at which a nerve impulse is relayed from the terminal portion of an axon to the dendrites of an adjacent neuron.

SYNAPSID A skull type, found in extinct mammal-like reptiles, possessing a single, low temporal opening.

SYNCOPE In reference to the effects of certain venomous SNAKE BITES, the collapse and/or unconsciousness of a bite victim, normally as a result of a sudden drop in blood pressure.

SYNDACTYL An animal possessing hands, or feet, in which two or more of the digits are growing fused together, as in the feet of chameleons (CHAMAELEONTIDAE).

SYNONYM 1. A term having the same or nearly the same meaning as another term, such as 'thoracic' and 'pectoral' **2.** A SCIENTIFIC NAME of a taxon no longer considered valid and rejected, or succeeded by another.

SYSTEMATICS; TAXONOMY The area of biology concerned with the study of the classification and diversity of reptiles and amphibians (and all other living organisms), and the natural associations and interactions between them.

SYSTEMIC Affecting the entire body, i.e., the cardiovascular, nervous, and respiratory systems.

T

TA Abbreviation for TACTILE ALIGNMENT.

TACHYCARDIA In reference to the symptoms of certain venomous SNAKE BITES, an unusually fast heart rate.

TACHYPHYLAXIS The rapid reduction in reaction to a dose of a toxic substance following previous injections of small doses of the same substance.

TACTILE Pertaining to the sense of touch.

TACTILE ALIGNMENT The form of behaviour in courting snakes, in which the male attempts to position his tail with that of the female so as to bring their cloacae into line. The cloacal or ANAL SPURS of male boids may be used to assist alignment, which precedes INTROMISSION and COITUS.

TACTILE BRISTLE; APICAL BRISTLE; TACTILE HAIR In certain lizards a single, short, sensitive filamentary 'hair' situated in the centre of the APICAL PIT on the dorsal scale.

TACTILE CHASE The form of behaviour in courting snakes in which the male follows, and attempts to mate with, the female. The male accompanies or closely follows the female, his tongue frequently flicking over her as she proceeds forward. He becomes excited as he tries repeatedly to locate her cloaca with his own (TACTILE ALIGNMENT) and his movements become spasmodic. Should the female show little or no interest and attempt to escape the male's attentions, he may try to restrain her with his jaws. If she is receptive, the male will follow up the chase with a TAIL-SEARCH COPULATORY ATTEMPT.

TACTILE HAIR *See* TACTILE BRISTLE.

TADPOLE The larva of an anuran, although the term is sometimes used to include the larvae of urodeles. A tadpole is characterized by a rudder-like, finned tail and, in the early days, external gills, on an egg-shaped body.

TAG Acronym denoting the Taxon Advisory Group (AAZPA).

TAIL; INTROMITTENT ORGAN In male frogs of the genus *Ascaphus* the 'tail' is an organ used in underwater copulation enabling the frog to fertilize the female internally thus prevent sperm from being lost in the fast-flowing rivers and streams that they inhabit.

TAIL CREST The cutaneous ridge on the upper surface of the tail of certain lizards, e.g., *Basiliscus plumifrons*, anuran larvae and many breeding male urodeles.

TAIL CURLING A reaction in some lizard species, e.g., *Leiocephalus carinatus*, to various external stimuli, in which the tail is rolled or curled vertically up and down.

TAIL-DROPPING The automatic defensive habit of many lizards in which the tail breaks off when seized (AUTOTOMY).

TAIL FILAMENT A fine, short, bristle-like appendage projecting from the centre of the usually truncated tail tip in the male palmate newt *Triturus helveticus* during the breeding season.

TAIL GLAND; ANAL GLAND The usually paired sac-like structure within the base of the tail in many reptiles, producing an odorous liquid used mainly in defence.

TAIL LENGTH In snakes and lizards, the distance from the rear margin of the ANAL PLATE to the tip of the tail.

TAIL-LESS AMPHIBIAN *See* ANURA.

TAIL-SEARCH COPULATORY ATTEMPT The form of behaviour in courting snakes in which the male attempts to locate the female's tail with his own in order to unite their cloacae. He coils his tail once or twice around her tail, moving the coils back and forth along its length so as to bring their cloacae into line. If the female is receptive she may elevate her tail in response to this action. If she is unreceptive, however, the male may raise her tail with his before giving up the attempt. The tail-search usually follows the TACTILE CHASE and precedes TACTILE ALIGNMENT.

TAIL SPINE 1. The NAIL on the tail tip of certain chelonians, e.g., *Testudo hermanni*.
2. The terminal scale on the tail tip of snakes which, in most species, is quite long and sharp but, in others, short and blunt. 3. The prominent thorn-like vertical protuberance on the base of the tail of male spine-tailed salamanders (genus *Mertensiella*), tactile in function and used to stimulate the female during mating.

TAIL TREMOR The quivering and wriggling of the tail in certain lizards, especially geckos, when excited, as for example when stalking insects.

TAIL-VIBRATING The rapid shaking of the tail tip in many snake species (e.g., *Pituophis, Elaphe*) when aroused or disturbed, producing a loud buzzing or rattling sound when the reptiles are amongst dried leaves or other debris.

TAIL WALK In some lungless salamanders (PLETHODONTIDAE), a stage in the sequence of behaviour displayed during courtship in which the female walks behind the male in response to the constant flicking and waving of his sharply bent tail.

TAIL-WAVING In a number of juvenile snakes, especially those of the arboreal pit vipers (e.g., *Trimeresurus*), the raising and waving around of the often bright and contrastingly coloured tail tip, which attracts inquisitive prey to within striking distance.

TAIPAN The name for either of two species, *Oxyuranus scutellatus* of New Guinea and extreme northern and North East Australia, and *Oxyuranus microlepidotus* of Central Australia, of the family ELAPIDAE. *Oxyuranus scutellatus* is the largest and most dangerous snake in Australia, attaining some 400 centimetres in length and found in a variety of habitats in Queensland and the northernmost parts of the Northern Territory.

TAPEWORM *See* CESTODA.

TAR Abbreviation for Thermo-activity range.

TARSAL **1**. Any one of the bones in the distal region of the hind limb between the TIBIA and/or FIBULA and the METATARSAL BONES in tetrapods. In anurans, those bones are greatly elongated and form the section of the hind limb between the shin and the foot. **2.** A scale in reptiles, situated between the digits and the joints of the tibia and/or fibula (ankle) of the hind foot.

TARSAL FOLD; TARSAL RIDGE A prominent fold or ridge of skin along the margin of the tarsal region of the foot in many anurans.

TARSAL SPUR An outgrowth on the heel.

TARSAL RIDGE *See* TARSAL FOLD.

TARSO-METATARSAL ARTICULATION The joint between the METATARSUS and the TARSUS.

TARSUS The bones (astragalus and calcaneum in anurans) of the hind limb which form the ankle.

TAXIS The progress of an animal in the direction of, or away from, a cause of stimulation in reaction to the strength and direction of the stimulus.

TAXON (pl.) **TAXA** A group, or rank, in TAXONOMY. The taxa

Bitis, Viperidae, Squamata, Reptilia and Chordata are examples of a genus, family, order, class and phylum respectively.

TAXONONMIC SPECIES Said of animals which have been compared on the basis of random sets of characteristics such as, for example, colour, scale arrangement or shape of appendages, for no other reason than that these are convenient and easily observed. Such features are used to describe species in field guides and in the formation of identification keys. *See* BIOLOGICAL SPECIES; EVOLUTIONARY SPECIES.

TAXONOMY; SYSTEMATICS The branch of biology concerned with the scientific classification of living things into groups, based upon all that is known or suspected of the evolution of the organisms concerned and on anatomical similarities etc.

TC Abbreviation for TACTILE CHASE.

Tdd Abbreviation for TRICOLOUR DIAD.

TD Abbreviation for TRICOLOUR DYAD.

TEBO *See* WAXWORM.

TECHNOPHAGY The act, witnessed occasionally in some reptiles, in which the female eats her own eggs. It has been observed in skinks of the genus *Eumeces* and in certain snakes which may devour unfertilised ova. The exact reasons for this behaviour are uncertain but they could include removing odours which would otherwise attract predators, or utilizing a valuable source of energy or, in captive reptiles, be the result of STRESS and/or suboptimal temperatures etc.

TEETH Hard, and in many species, dense structures situated on the jaws and/or over the palate in reptiles and amphibians, used primarily for biting, seizing and holding prey and, in some species, for partially masticating food. In many snakes, and one family of lizards (HELODERMATIDAE), some of the teeth, the fangs, have been modified for the conduction of venom. Depending upon the way in which the teeth are attached to the jaw, their various types include ACRODONT, THECODONT and PLEURODONT. Those which possess some form of venom-conducting groove include AGLYPHIC, OPISTHOGLYPHIC, PROTEROGLYPHIC and SOLENOGLYPHIC TEETH.

TEIIDAE Whip-tails, racerunners, tegus and allies. Family of the SAURIA, inhabiting North and South America and the West Indies. Approximately 130 species in some nine genera.

TEMPERATE Having, or experiencing, a well-defined seasonal variation in temperature; a region, or any species from that region, which experiences mild or moderate climates intermediate to those of the polar and tropical regions.

TEMPERATURE-DEPENDENT SEX DETERMINATION The phenomenon in some crocodilians, chelonians and lizards, in which the sex of the eventual hatchlings may be determined by the temperature at which the eggs are incubated. Sex in snakes is generally determined genetically rather than by temperatures experienced during incubation.

TEMPERATURE GRADIENT A particular range in the temperature required by a reptile for it to regulate its internal body temperature.

TEMPERATURE-INDUCED PATTERN ABERRANCY The phenomenon in certain GRAVID female snakes which, when exposed to temperatures slightly above or, more commonly, below the recommended gestation temperature, produce young with varying degrees of pattern abnormality.

TEMPERATURE-SENSITIVE ORGAN *See* LABIAL PIT; LOREAL PIT; PIT.

TEMPEROLABIAL; TEMPEROLABIAL SCALE A large scale which projects downwards between the fifth and sixth SUPRALABIAL scales. Most Australasian elapids possess a temperolabial, including the dangerous species, but it is notably absent from the eastern brownsnake *Pseudonaja textilis*.

TEMPORAL Of, on, or relating to, the region of the temple on the side of the head, e.g., any one of the more or less enlarged scales situated behind the POSTOCULAR scales, beneath the PARIETAL and above the upper LABIAL scales at the angle of the jaw in snakes and lizards.

TEMPORAL HORN Any one of the large, often long, hornlike spines on the posterior margin and sides of the head in the horned lizards (*Phrynosoma*)

TERATOGENIC Causing birth defects or malformations. Teratogenesis in captive reptiles may result from a number of factors, including the use of X-rays in determining whether a specimen is GRAVID, or incorrect temperatures during gestation or egg incubation.

TERCIOPELO *See* BARBA AMARILLA.

TERENT'EV, **Pavel Victorovich,** (1903-1970), Russian herpetologist, professor of vertebrate zoology at Leningrad University. Contributed over 150 scientific works, including the *Key to the reptiles and amphibians of the USSR*, co-authored with S. A. Chernov (published in Russian in 1949 and translated into English in 1965),*The frog* (1950 in Russian), and *Herpetology - a manual on amphibians and reptiles* (published in Russian in 1961 and translated into English in 1965).

TERMINAL PHALANX The distal bone in each of the digits of the forefoot (hand) or hind foot.

TERRACED Descriptive of the prominent ridge situated below the lip in some lizards of the AGAMIDAE, resulting from the difference, in both shape and size, between the inner and outermost scales of the chin.

TERRAPIN Any one of various semi-aquatic freshwater chelonians, forming the family EMYDIDAE.

TERRARIUM (pl.) **TERRARIA** A generally closed container of any shape or size and usually furnished with plants, in which amphibians, reptiles or other small animals are housed.

TERRESTRIAL Living upon the land, as opposed to the sea or air; dwelling largely at ground level.

TERRITORIAL BEHAVIOUR Any type of action or response associated with the establishment, possession or defence of an area from which rivals are driven by means of visual threats and/or ritualized combat. Such behaviour occurs in many reptiles and amphibians, e.g., dendrobatid frogs, crocodilians and many lizard species.

TERTIARY The larger and older geological period of the CENOZOIC era, following the CRETACEOUS, some 65 million years ago, and extending to the beginning of the QUATERNARY, about 2 million years ago. The Tertiary is made up of the Palaeocene, Eocene, Oligocene, Miocene and Pliocene epochs, and characterized by the emergence of modern mammals.

TESTICULAR REGRESSION *See* REFRACTORY PERIOD.

TESTIS (pl.) **TESTES** The male reproductive organ in which spermatozoa are produced. Reptiles and amphibians have two testes, as do all vertebrates.

TESTUDINES *See* CHELONIA.

TESTUDINIDAE Tortoises. Family of the CHELONIA, suborder Cryptodira, inhabiting America, Europe, Africa and Asia. Not found in Australia. 41 species in 9 genera.

TETRADACTYL Having four digits, as in the hind foot of a crocodilian for example.

TETRAPOD Any four-limbed vertebrate animal including reptiles, amphibians, birds and mammals. The limb skeleton of all tetrapods is based upon the PENTADACTYL LIMB of five digits.

TEXTILOTOXIN A component of the venom from the Australian elapid *Pseudonaja textilis*, the brown snake and the most potent NEUROTOXIN known, requiring only one millionth of a gram to kill a mouse.

THANATOSIS; DEATH FEINT; LETISIMULATION The act of playing dead, in which the locomotory muscles become totally relaxed. Thanatosis occurs in a number of snakes and anurans when threatened, disturbed or manhandled.

Th-Ar Abbreviation for THERMO-ACTIVITY RANGE.

THECA A membranous cup-shaped envelope or sheath. The term has been used in herpetology for the BROOD POUCH of marsupial frogs (*Gastrotheca*).

THECODONT Possessing teeth each located within a bony socket, as in crocodilians.

THERMAL BURN In general, any burn resulting from body contact with a naked (unprotected) heat source. In reptiles and amphibians such burns usually occur on the ventral surface of the animal.

THERMAL CONFORMIST An ECTOTHERM that allows the temperature of its body to vary with that of its surroundings. *Contrast* THERMAL NONCONFORMIST

THERMAL NONCONFORMIST An ECTOTHERM that endeavours to reach and sustain a regular body temperature that may or may not be the same as its surroundings. *Contrast* THERMAL CONFORMIST.

THERMORECEPTOR Any device, or organ, that is sensitive to heat or variations in temperature, such as the LABIAL PIT of many boid snakes and the PIT of crotaline vipers.

THERMOREGULATION The maintenance of the OPTIMAL TEMPERATURE RANGE in POIKILOTHERMS. In reptiles and amphibians, which are unable to produce metabolic heat to raise their body temperature, thermoregulation must take place by basking when too cool and seeking shade when too warm, in order that bodily functions such as respiration, digestion, shedding etc., can continue normally.

THIGMOTHERM A species that thermoregulates by seeking direct contact with a preheated surface such as a rock or road surface for example. Many nocturnal snakes are thigmothermic whilst many diurnal snake and lizard species employ a combination of thigmothermic and heliothermic strategies.

THIRD EYE *See* PARIETAL EYE; PINEAL COMPLEX.

THIRD EYELID *See* NICTITATING MEMBRANE.

THORACIC Of, on, or relating to, the anterior region of the body trunk of vertebrates containing the heart and lungs within the rib cage (the thorax).

THORACIC LAMINA; PECTORAL LAMINA Either one of the third pair of horny epidermal scutes on a chelonian PLASTRON.

THORN The terminal spine-like scale on the tail tip of blind snakes (TYPHLOPIDAE).

THORNY DEVIL *See* MOLOCH.

THREADWORM *See* OXYURIDAE

THREAT BEHAVIOUR Any type of innate, species-specific action or response associated with aggressive, intra- and interspecific combat over territory, females, food etc., and used frequently in the intimidation of potential enemies, predators and aggressors. Threat behaviour amongst reptiles and amphibians occurs in many forms and includes flattening the body, inflating the neck or the whole body, opening the mouth to reveal vivid and brightly coloured interiors, raising and stiffening the legs and arching the back, hissing, biting and displaying brightly coloured parts of the body.

THROAT FAN The DEWLAP on the throat in several lizard genera, especially of the family IGUANIDAE.

THROATS In the commercial skin-trade, the V-shaped pieces of skin taken from the sides of the neck and beneath the chin of very large adult caimans (mainly *Melanosuchus*, occasionally *Caiman*).

THROMBIN An enzyme that acts on FIBRINOGEN in blood causing it to clot.

THROMBOCYTOPENIA Showing a shortage in the successful number of circulating blood platelets.

THROMBOSIS In reference to the symptoms of particular venomous SNAKE BITES, the state in which the blood coagulates and clots within a blood vessel or inside the heart and remains at the place of its creation often obstructing the flow of blood and a frequently fatal consequence of specific snake venoms.

THUMB PAD A NUPTIAL PAD, situated at the base of the first, or innermost, digit of the forefoot in some male anurans.

THYROID DISEASE *See* GOITRE.

TIBIA The large innermost bone of the lower hind limb of tetrapod vertebrates, extending from the knee to the heel or, in anurans, the corresponding part of the leg.

TIBIAL Any one of the scales situated on the hind limb of a lizard, between the knee and the heel (ankle).

TIBIAL GLAND A large, swollen gland on the upper surface of the lower hind limb (TIBIA) in some toads of the genus *Bufo*.

TIBIOFIBULA The bone of the lower hind limb in amphibians, formed from the fusion of the TIBIA and the FIBULA.

TIBIO-TARSAL ARTICULATION The joint in the hind limb that allows movement between the TIBIA and the TARSUS.

TIC POLONGA *See* DABOIA.

TICH POLONGA *See* DABOIA.

TICK A blood-sucking ECTOPARASITE of the order

Acarina (suborder Ixodides) and the largest of the MITES, some species reaching as much as 3 cm when fully fed. Ticks have worldwide distribution and over 300 different species have so far been described. So-called 'hard ticks' (Ixodidae) have a toughened plate above a forward-directed head. 'Soft ticks' (Argasidae), on the other hand, possess a rubbery integument, lack a head plate, and the head is situated on the underside. Adult ticks possess four pairs of legs whereas young ticks have only three pairs. Most wait on the tips of foliage and grasses with outstretched forelegs ready to attach themselves to any creature passing by, falling from their HOST after feeding. Some ticks are vectors of disease and certain species can cause mild paralysis in man. Ticks are frequently found infesting the skin of reptiles, especially snakes, chelonians and varanid lizards.

TL Abbreviation for TOTAL LENGTH.

TM Abbreviation for TRICOLOUR MONAD.

TOADFLY A fly (*Bufolucilia bufonivora*) which deposits its eggs within the nasal openings of bufonid toads The fly larvae feed mainly upon mucous but, as they grow, they frequently penetrate into the eyes and brain of the toad with usually fatal consequences. Several other fly species have larvae which can develop within other anurans.

TOADLET A newly metamorphosed toad.

TOE-CLIPPING A method of marking lizards by clipping one or more toes in a predetermined combination in order that individuals can be recognised both in the field and in the laboratory.

TOE DISC *See* DISC.

TOKEN AMPLEXUS A brief, somewhat ritualized, form of AMPLEXUS in certain anurans which deposit their eggs in situations other than in water, e.g., *Mantella aurantiaca*, *Rhinoderma darwini* and several *Dendrobates* species.

TOMIGOFF *See* BARBA AMARILLA.

TONGUE-FLICKING The protrusion and, in some species, the up-and-down movement of the tongue, in many snakes and lizards, during which scent particles adhere to the moist upper surface and are transferred to the JACOBSON'S ORGAN in the roof of the mouth for analysis. In snakes, the tongue is extended for a few seconds while the mouth is closed, by means of a V-shaped indentation (the LINGUAL FOSSA) in the underside of the ROSTRAL scale, and the flicking increases during the pursuit of prey, or a mate, or when the reptile is agitated or annoyed. In certain lizards, such as the blue-tongued skinks (*Tiliqua*), the brightly coloured tongue is protruded for several seconds, sometimes through an open mouth, as a form of threat behaviour towards an aggressor.

TONGUE WORM A modified worm-like arachnid that, either as a larva or a 12 centimetre-long adult, inhabits the respiratory tract of amphibians and reptiles, especially crocodilians.

TOOTH REPLACEMENT

TOPOGRAPHY Description of the external features of a reptile or amphibian.

TOP PREDATOR *See* FOOD CHAIN.

TORPOR An energy-saving period of inactivity, in which bodily functions are greatly slowed, exhibited by many reptiles and amphibians during adverse climatic conditions. *See* AESTIVATION; HIBERNATION.

TORTOISESHELL; CAREY The horny, semi-transparent, yellow and dark brown mottled laminae from the shell of the marine hawksbill turtle *Eretmochelys imbricate*

TOTAL LENGTH The greatest, combined, straight-line length of a reptile (with the exception of chelonians), caecilian or tailed amphibian, measured from the tip of the snout to the tip of the tail. *See* HEAD-BODY LENGTH; SNOUT-VENT LENGTH.

TOXIC Of, relating to, or caused by a TOXIN or poison; poisonous; harmful; or deadly.

TOXICITY The noxiousness of a toxin.

TOXICOLOGY The branch of science concerned with the study of naturally occurring poisonous substances (toxins), their nature and effects upon living organisms, and antidotes effective against them. *Compare* TOXINOLOGY.

TOXIN *See* TOXICOLOGY.

TOXINOLOGY A subdivision of TOXICOLOGY that is concerned specifically with the toxins produced by organisms (e.g., the venom of snakes and the poisonous secretions of amphibians). *Compare* TOXICOLOGY.

TP Abbreviation for TRICOLOUR PENTAD.

TRACHEA The membranous tube, or windpipe, in air-breathing vertebrates, stiffened by incomplete cartilaginous rings and leading from the throat to the lungs allowing the passage of air. In many snakes the anterior part of the trachea is thrust forwards on the floor of the mouth to allow the reptile to breathe whilst in the process of swallowing large prey items.

TRACHEAL Pertaining to the TRACHEA.

TRACHYSTOMATA *See* SIRENIDAE. *See* AMPHIBIAN.

TRACTABLE Said of any reptile or amphibian that is easily controlled or managed; inoffensive.

TRAFFIC Acronym denoting the Trade Records Analysis of Flora and Fauna in Commerce. Founded in the mid 1970s to ensure that trade in wild plants and animals does not become a threat to the conservation of nature.

TRAMP SPECIES A species that has been accidentally spread around the world by the actions of human commerce.

TRANSECT To cut or divide crossways. In ecology, a line marked within an area being surveyed in order to provide a way of measuring and demonstrating geographically the distribution of particular species, especially when they are arranged in a linear sequence (e.g., across a woodland margin, or up a riverbank). *See* BELT TRANSECT and LINE TRANSECT.

TRANSFORMATION A noticeable and usually relatively sudden change (METAMORPHOSIS) in the shape and structure (and usually also in the habits) of most amphibians as they progress from larval to adult form, e.g., tadpole to frog.

TRANSILLUMINATION The technique, commonly called 'candling', used to determine the viability of eggs, through which a strong light is directed briefly in a darkened room during the early stages of incubation. Viable eggs clearly show blood vessels which appear as a web-like network of thin red lines. An absence of these blood vessels indicates that the embryo has died or that the eggs have not been fertilized.

TRANSLOCATION The moving of reptiles and/or amphibians from one area to another and releasing them there. In southern England, for example, colonies of the sand lizard *Lacerta agilis* and the smooth snake *Coronella austriaca*, have been translocated from districts where they were under threat to new, less disturbed areas.

TRANSVERSE Crossing diagonally; from side to side.

TRANSVERSELY DIVIDED Pertaining to a scale, on the head of some crotaline vipers, that is divided across its long axis to form two smaller scales, e.g., the first INFRALABIAL in *Crotalus ruber* and the upper PREOCULAR in *Crotalus lepidus*.

TRANSVERSE VENTRAL SCALE ROW The arrangement of the ventral scales in crocodilians, situated in transverse rows, with the number of rows between the neck and vent varying from one species to another. The rows are counted from that directly behind the collar up to, but excluding, the row bordering the vent.

TREE BARK TUMOUR *See* POX.

TREE CANOPY *See* CANOPY.

TREMATODA Class of parasitic PLATYHELMINTHES known as flukes, that attach themselves to a HOST by means of hooks and suction discs. They occur throughout the alimentary canal and other internal organs of both amphibians and reptiles.

TRIAD A group of three. A term commonly used for the trio of black rings in the pattern of some coral snakes (*Micrurus*) and their mimics (*Erythrolamprus*), in which successive arrangements of black-yellow (or white)-black-yellow (or white)-black rings are isolated from the others by red rings.

TRIASSIC The oldest period of the MESOZOIC era, beginning some 230 million years ago at the end of the PERMIAN and extending for about 35 million years until it was succeeded by the JURASSIC. It is characterized by an increase in the number of primitive amphibians and reptiles, including phytosaurs, dinosaurs and chelonians.

TRICARINATE Descriptive of a scale, lamina or chelonian carapace bearing three ridges or keels.

TRICOLOUR A convenient grouping term for any of the snakes of the genus *Lampropeltis* which have patterns of three alternating rings or bands of black, red and yellow (or white). Includes most of the *L. triangulum* group, commonly known as 'milk snakes'. The term has been used occasionally by some authors for certain coral snakes of the genus *Micrurus*.

TRICOLOUR DIAD A pattern in certain snakes characterized by red rings or bands, each isolated from the other by a group of alternating rings or bands of black-light-black, i.e., two black between two red.

TRICOLOUR MONAD A pattern in certain snakes characterized by red rings or bands, each isolated from the other by a group of alternating rings or bands of light-black-light, i.e., one black between two red.

TRICOLOUR PENTAD A pattern in certain snakes characterized by red rings or bands, each isolated from the other by a group of alternating rings or bands of black-light-black-light-black-light-black-light-black. i.e., five black between two red.

TRICOLOUR TETRAD A pattern in certain snakes characterized by red rings or bands, each isolated from the other by a group of alternating rings or bands of black-light-black-light-black-light-black, i.e., four black between two red.

TRICOLOUR TRIAD A pattern in certain snakes characterized by red rings or bands, each isolated from the other by a group of alternating rings or bands of black-light-black-light-black, i.e., three black between two red.

TRICUSPID Possessing three tooth-like projections or CUSPS.

TRIDACTYL Possessing three digits on the fore- or hind foot.

TRINOMEN *See* TRINOMIAL NAME.

TRINOMIAL NAME The three-part scientific name (trinomen) of a reptile or amphibian that incorporates its genus, species and subspecies, e.g., *Pantherophis obsoleta quadrivittata* (a subspecies of the American rat snake, *P. obsoleta*). The system of giving a scientific name to a subspecies in which a third name is added to the genus and species is called trinomial nomenclature. *See* BINOMIAL NOMENCLATURE.

TRINOMIAL NOMENCLATURE *See* TRINOMIAL NAME.

TRIONYCHIDAE Soft-shelled turtles. Family of the CHELONIA, suborder Cryptodira, inhabiting most temperate and tropical regions of the world with the exception of Central and South America, Madagascar and Australia. 28 species in six genera.

TRISMUS In reference to the effects of certain venomous snake bites, an inability to open the mouth due to uninterrupted contractions of the jaw muscles and caused by a form of tetanus, or lockjaw.

TROGLODYTIC Cave-dwelling. Although many reptiles and

amphibians venture into caves either by accident or in pursuit of prey, only a few urodeles are true troglodytes, characterized by their loss of pigmentation and vision, e.g., the olm *Proteus anguineus* and several members of the genus *Eurycea*.

TROGONOPHIDAE Family of the SQUAMATA, suborder Amphisbaenia, inhabiting north-western Africa, Somalia, the Arabian peninsula and Socotra Island in the Indian Ocean. Six species in four genera.

TROPICAL Of, relating to, or inhabiting, the tropics; occurring in regions between the latitudes of 23½° N (tropic of Cancer) and 23½° S (tropic of Capricorn) of the equator.

TROPICAL ARID FOREST A sparse vegetation type, occurring at elevations of between 100 and 600 metres on the leeward side of high ground, in the path of rain-bearing winds, receiving normally no more than 1000 millimetres of rain annually, and characterized by scattered cacti, stunted trees and scrub, and several types of grasses.

TROPICAL DECIDUOUS FOREST A vegetation type, typical of low-lying, semi-arid regions of the tropics, with trees and shrubs that shed their foliage in the dry season.

TROPICAL EVERGREEN FOREST A vegetation type, typical of low-lying, semi-humid regions of the tropics, with trees and shrubs that bear foliage throughout the year.

TROPICOPOLITAN Confined in distribution to regions within the tropics. *Contrast* COSMOPOLITAN.

TROPIDURIDAE Neotropical ground lizards. FAMILY of the SQUAMATA, suborder SAURIA, formerly of the IGUANIDAE, inhabiting South America, the Galapagos Islands and the West Indies. Some 29 species in six genera.

TRYPANOSOME Any one of a number of parasitic protozoans of the genus *Trypanosoma*, found in the blood and intestines of amphibians and reptiles and transmitted by certain insects (Diptera) and possibly mites (ACARINA). Leeches (Hirundinea) are largely responsible for the transmission of trypanosomes in aquatic reptiles and amphibians.

TSCA Abbreviation for TAIL-SEARCH COPULATORY ATTEMPT.

TSD Abbreviation for TEMPERATURE-DEPENDENT SEX DETERMINATION.

TUATARA A lizard-like reptile of the order RHYNCHOCEPHALIA, inhabiting certain small islands off New Zealand, and the sole surviving representative of a group common during the MESOZOIC. There are just two species, *Sphenodon punctatus* and *Sphenodon guentheri*.

TUBERCLE A small knoblike projection or elevation on the skin, e.g., the PALMAR TUBERCLE on the ventral surface, or palm, of the forefoot in anurans.

TUBERCULATE Having small, raised, rounded nodules on the skin; covered with tubercles.

TUBERCULOSIS A communicable bacterial disease caused by *Mycobacterium* micro-organisms, occasionally affecting the respiratory or intestinal system or, less commonly, the skin of reptiles and amphibians. Usually occurs as a secondary disease in animals already weakened by other conditions, e.g., suboptimal husbandry, poor nutrition, injury etc.

TUMOUR Any abnormal swelling or tissue growth forming as a result of the uncontrollable development of new cells, occurring occasionally on the skin, bones or internal organs of reptiles and amphibians. Such tumours may be malignant (e.g., thyroid carcinoma, black melanosarcoma) or, more commonly, benign (e.g., papillomata).

TUNDRA An immense treeless zone extending between the ice cap and the timber line of Eurasia and North America with a permanently frozen subsoil as a result of the prevailing climate, allowing only minimum vegetal growth in the form of dwarf shrubs and bryophytes. No reptiles or amphibians occur within the tundra.

TUSK *See* DENTARY PSEUDO-TEETH.

TWIN-EGG Two eggs which have adhered together in the oviduct of a reptile. Such an egg may be the cause of dystocia, or EGG-BINDING.

TWINNING The presence of two embryos within the same egg, resulting from two separate embryos being unintentionally enclosed in the same shell during their progress through the oviduct.

TYMPANIC Of, or relating to, the region of the TYMPANUM; a scale situated directly above the tympanum in some lizards.

TYMPANIC DISC *See* TYMPANUM.

TYMPANIC MEMBRANE *See* TYMPANUM.

TYMPANIC SHIELD *See* TYMPANUM.

TYMPANUM; EARDRUM; TYMPANIC DISC; TYMPANIC MEMBRANE; TYMPANIC SHIELD The membrane separating the middle ear from the outer ear, vibrating in response to sound waves and communicating them, by means of the OSSICLE of the middle ear, to the site of hearing. In many amphibian and reptile forms the tympanum is exposed at the skin surface, while in others it may be hidden or completely lacking. It is particularly well-developed in many anurans and often larger in males than in females.

TYPE The original specimen used for naming and describing a species or subspecies. If this specimen is the one collected by the author who first published a description of, and gave a scientific name to, the species it is termed a 'holotype'. The place of origin of the type specimen is termed the 'type locality'.

TYPE GENUS The genus chosen as a standard of reference for a family. For example, the lizard genus

Lacerta is the type genus of the family LACERTIDAE.

TYPE LOCALITY *See* TYPE.

TYPE MATERIAL A shared term for all TYPE specimens collected from the wild. Such specimens should always be quickly moved to public institutions where they will be both safe and available to other researchers.

TYPE SPECIMEN *See* TYPE.

TYPHLONECTIDAE Aquatic caecilians. Family of the APODA, inhabiting tropical and subtropical South America. Six species in five genera.

TYPHLOPIDAE Blind snakes. Family of the SQUAMATA, suborder SERPENTES, infraorder Scolecophidia, inhabiting most of the warmer parts of the world. Over 180 species in some four genera.

TYPICAL RACE A subspecies that bears a specific name identical to that of the species, e.g., *Epicrates cenchria cenchria* (the rainbow boa) or *Chelydra serpentina serpentina* (the snapping turtle).

U

ULCERATION The growth or development of a lesion that opens on the skin's, or an organ's, surface.

ULCERATIVE GINGIVITIS *See* STOMATITIS.

ULCERATIVE STOMATITIS *See* STOMATITIS.

ULNA One of a pair of bones in the forelimb of TETRAPODA, or four-limbed animals.

ULNAR 1. Located to the rear of the body axis. **2**. In relation to the hindmost part of a vertebrate limb, often in reference to anurans.

ULTRAGULOSITY A term, used in old literature, meaning extraordinary greediness; used in reference to the feeding habits of certain anurans.

ULTRASOUND-SCANNING High-frequency sound waves given off by a transducer which is moved over an animal's body to determine if it is pregnant, or GRAVID. The presence of the comparatively dense ovarian FOLLICLES is determined by the pattern of reflected sound waves. Ultrasound is completely harmless to the developing embryo, unlike X-rays which can produce certain CONGENITAL conditions.

ULTRA VIOLET *See* ULTRAVIOLET RADIATION.

ULTRAVIOLET RADIATION Lying beyond the violet end of the visible spectrum, ultraviolet is important in the successful husbandry of many reptiles, particularly desert-living tortoises and lizards, as it is necessary for calcium metabolism and the stimulation of pigment cells and sex glands. It is also essential for vitamin D synthesis. Reptile keepers can obtain most of the ultraviolet radiation needed from various forms of mercury-vapour lamps. Energy in ultraviolet radiation is generally within the range of about 100-380 nanometres (nm) but the UV band is broken down further like so: ozone-producing (180 - 220 nm); bactericidal (220 - 300 nm); erythemal, resulting in reddening of the skin (280 - 320 nm); and 'black light' (320 - 400 nm). The International Commission on Illumination (CIE) identifies three UV bands: UV-A (315-400 nm) visible to the naked eye and responsible in captive reptiles for inducing normal behaviour such as feeding and mating; UV-B (280 - 315 nm) invisible to the naked eye and responsible for sun tans in humans. It allows for synthesis of vitamin D3 which allows reptiles to process calcium thus averting bone disease; and UV-C (100-280 nm) invisible to the naked eye and employed in sterilization, killing bacteria. It is extremely dangerous and can result in damage to DNA.

UMBO The slightly convex or rounded hump in the centre of an individual LAMINA of a juvenile chelonian PLASTRON.

UNAVAILABLE NAME In nomenclature, any name which does

not satisfy the ICZN's strict requirements. *Compare* AVAILABLE NAME *and See* VALID NAME

UNCIFORM The small bone in an anuran hand, formed into a single component by the union of the fourth and fifth CARPAL BONES.

UNDULATING MOTION; UNDULATORY MOVEMENT *See* SERPENTINE MOVEMENT.

UNESCO Acronym denoting the United Nations Educational, Scientific, and Cultural Organisation. Founded on 16 November 1945 with the aim of promoting international cooperation among its 190 Member States and six Associate Members in the fields of education, science, culture and communication.

UNGUAL Referring to or related to a CLAW; possessing claws.

UNGUIS A claw, or the part of the DIGIT from which it develops.

UNICUSPID Possessing a single CUSP.

UNISERIAL; UNISERIATE Arranged in a single row.

UNISERIATE *See* UNISERIAL.

UNKEN REFLEX Particular actions employed by certain amphibians, e.g., *Bombina* and *Dendrophryniscus*, in response to threat or disturbance and which form an APOSEMATIC display. In the fire-bellied toad *Bombina orientalis* the back is flattened and arched, with the head held up and back, and the palms of the hands and feet are raised and reversed in a warning attitude, exposing the vivid red-and-black flash coloring on the ventral surface and generally accompanied by an increase in skin secretion.

UPPER LABIAL *See* LABIAL

UREA Nitrogen-containing substance excreted by some animals, e.g., fish, amphibians and mammals.

UREOTELIC Excreting nitrogen in the form of UREA. Amphibians are ureotelic.

UREOTELISM The condition, found in terrestrial chelonians and amphibians that dwell in damp situations, in which UREA is the main product of kidney excretion.

URETER One of a pair of ducts occurring in reptiles (and birds and mammals) that carries URINE to the CLOACA from the kidneys. Linked with the METANEPHROS, it replaces the WOLFFIAN DUCT occurring in amphibians.

URIC ACID A white, odourless and virtually insoluble crystalline product of protein metabolism, transformed from nitrogenous waste by animals that inhabit arid areas. The substance, in terrestrial reptiles, that is produced from the breakdown of purine and which is the most significant form in which metabolic nitrogen is excreted.

URICOTELIC Relating to those animals that excrete URIC ACID rather than UREA and characteristic of many terrestrial animals in which the conservation of water is essential. Reptiles are uricotelic.

URINE The slightly acid, fluid discharged through the CLOACA. It is produced in the kidneys and contains UREA (in amphibians) or URIC ACID (in reptiles), and many other substances in small amounts.

URODELA A recent order of tailed amphibians comprising the salamanders, newts, and related forms. Around 450 species in some eight families are dispersed chiefly in the northern temperate region although several genera extend over the equator into South America The majority are lizard-shaped with four limbs but some are elongated and eel-like with the limbs vestigial. Unlike the ANURA, the tail is not lost during METAMORPHOSIS and fertilization is internal via SPERMATOPHORES.. Sexual dimorphism is common, the males of many species having greatly enhanced colouration and an increase in the size of the median fin during the breeding season. *See* CAUDATA.

URODAEUM In reptiles, the central chamber of the CLOACA into which opens the ureter, bladder and, depending on the sex of the animal, the VASA DEFERENTIA or OVIDUCT.

UROPELTIDAE Shield-tailed snakes. Family of the SQUAMATA, suborder Henophidia, inhabiting southern India and Sri Lanka. Over 50 species in some eight genera.

UROSTEGE Any one of the enlarged scales on the ventral surface of the tail; SUBCAUDAL scales.

UROSTYLE A pointed rod of bone at the hind end of the vertebral column of frogs and toads. Greatly elongated, it reaches from the middle of the back to the posterior end of the body and is formed by the fusion of several caudal vertebrae.

URTICARIA In reference to the symptoms of certain venomous SNAKE BITES, a skin condition characterized by the formation of red or whitish raised patches, usually caused by an allergic reaction.

URUTU *Bothrops alternatus,* a PIT VIPER of the family VIPERIDAE, subfamily Crotalinae, inhabiting Brazil, Paraguay and northern Argentina. Responsible for a large number of bites each year.

USDA Acronym denoting the United States Department of Agriculture.

USDI Acronym denoting the United States Department of the Interior.

USFWS Acronym denoting the United States Fish and Wildlife Service. The Service undertakes to work with others to conserve, protect, and enhance fish, wildlife, plants, and their habitats for the continuing benefit of the American people.

UTERINE MILK A creamy, paste-like substance produced in the OVIDUCT wall of viviparous caecilians, consisting largely of emulsified fats and orally assimilated by the developing foetuses.

UV *See* An abbreviation for ULTRA VIOLET

UV-A *See* ULTRAVIOLET RADIATION.

UV-B *See* ULTRAVIOLET RADIATION.

UV-C *See* ULTRAVIOLET RADIATION.

UVR An abbreviation for ULTRAVIOLET RADIATION.

V

VACUITY A space between bones of the skull.

VAGILE A term applied to an animal that is at liberty to move about.

VAGILITY In reference to the speed and amount of movement of a species. A highly VAGILE species is able to move around freely, frequently, and over long distances.

VAGINA; VAGINAL POUCH In reptiles the entire and most caudal segment of the OVIDUCT adapted to receive the male INTROMITTENT ORGAN.

VAGINA DENTALIS The term for the sheath-like membrane of tissue that covers and protects the FANG of venomous snakes.

VALID NAME In nomenclature, the accurate name for a given TAXON, which may have several AVAILABLE NAMES one of which, typically the oldest, is selected as the valid name. The valid name is always an available name although an available name is not always a valid one. *Compare* INVALID NAME.

VARANIDAE Monitors. Family of the SQUAMATA, suborder SAURIA, inhabiting Africa, South-East Asia and Australasia. Approximately 30 species in a single genus (*Varanus*).

VARIATION Disparity exhibited by individuals within a species and which may be chosen or removed by natural selection. In sexual reproduction, gene restructuring in each generation guarantees the continuation of variation, the definitive basis of which is mutation which generates new genetic material.

VASCULAR SYSTEM The specialized network of veins, arteries and lymph vessels for the circulation of fluids throughout the body tissues. The blood-vascular system of animals enables the passage of nutrients, excretory products, respiratory gases and other metabolites into and out of the cells. The arterial system transports blood to different parts of the body from the heart, the venous system returns the blood to the heart, and the lymphatic system transports the fluid which permeates into the tissues from the blood.

VASCULITIS The inflammation of a vessel.

VAS DEFERENS (pl.) **VASA DERERENTIA** One of the main pair of ducts that carry sperm from the TESTES to the exterior.

VAS EFFERENS (pl.) **VASA EFFERENTIA** In reptiles, any of a number of various small ducts that carry sperm from the seminiferous tubules of the testis in the EPIDIDYMUS.

VASODILATOR Any agent which causes blood vessels to relax.

VASOPRESSOR Any one of a group of drugs that raises blood pressure by constricting the arteries.

VECTOR Any animal, such as a tick or mite, that can convey a disease-producing organism from an infected animal to a healthy one, either within or on the surface of the body.

VELDT The open grassy areas of the African highlands.

VELUM PALATI The upper of two muscular folds, or plicae, located on the palate in the rear of the mouth of a crocodilian. Its function, in combination with the lower fold (BASIHYAL VALVE), is to prevent water from entering the glottis when a submerged animal opens its mouth.

VENENATION *See* ENVENOMATION

VENOM The substance secreted by specialized glands of certain animals, such as venomous reptiles, that is capable of bringing about a toxic reaction when introduced into the tissues of another animal. A white, greenish or yellow viscous fluid, venom is a complicated mixture of substances. Dried, it consists of 90-95% various proteins with varying degrees of toxicity and effectiveness, including proteins with enzymatic features and non-toxic proteins. Snake venoms frequently contain small quantities of other organic compounds, e.g., carbohydrates, lipids and free amino acids, and metals such as potassium, manganese, sodium and calcium.

VENOM APPARATUS The structural mechanism that manufactures, conveys and dispenses the venom. In snakes it is usually made up of a pair of VENOM GLANDS, a pair of VENOM DUCTS and two or more teeth, or FANGS.

VENOM DUCT The tube that extends from the VENOM GLAND to the entrance of the FANG in venomous snakes. The duct passes beneath the eye, pit (if present) and nostril, and over the upper bone of the jaw (the MAXILLA).

VENOM GLAND A gland in reptiles, evolved from one of the salivary glands. Located between the eye and the angle of the mouth, the gland carries venom through the VENOM DUCT to the FANG which, depending upon the species concerned, is either grooved or hollow. In the HELODERMATIDAE (beaded lizards), the venom gland is the inferior labial salivary gland which has become specialized for the production of venom which is secreted through the mucus membrane of the lower jaw and collects at the base of the teeth.

VENOMOUS Relating to an animal that possesses VENOM. Until recently the term 'venomous' was used as an alternative to POISONOUS (e.g., a poisonous snake, venomous frog) but today the term 'poisonous' is normally used in reference to those animals which have a noxious or harmful effect on another when partially or completely devoured, and 'venomous' for animals that introduce venom into the body of another by means of specialized teeth (fangs) or a sting.

VENT The hind opening located on the under surface of the body at the beginning of the tail; the external entrance/exit of the CLOACA.

VENTILATION *See* ARTIFICIAL RESPIRATION.

VENTER The abdomen or belly of an animal; the complete undersurface.

VENT LOBES Fleshy lobes occurring in certain male urodeles, usually behind and on each side of the VENT.

VENTRAL Any one of the enlarged scales on the under surface, or VENTER, of a snake, from the head to the tip of the tail but generally used in reference to those ventral scales between the head and the ANAL PLATE.

VENTRAL COLLAR In the majority of crocodilians, a distinctive row of enlarged scales across the throat just anterior to the forelimbs. One species has two rows of enlarged scales, forming a double collar, and a few species are without the row of enlarged scales altogether, so their collar is not obvious

VENTRAL COUNT *See* SCALE COUNT.

VENTRAL KEEL A prominent ridge on the middle of the ventral scales of certain snakes, especially certain members of the HYDROPHIIDAE.

VENTRICLE A lower chamber of the heart, having thick, muscular walls and receiving blood from the upper chamber (atrium), which it pumps to the arteries. The hearts of reptiles have two ventricles but those of amphibians have just one.

VENTROLATERAL Relating to the part of the body where the side unites with the under surface.

VERMICIDE Any substance employed to kill and eliminate intestinal worms.

VERMICULATE Decorated with irregular or wavy tracery or markings resembling the tracks of worms.

VERMICULITE A form of expanding mica which is produced for the insulation of buildings, fireproofing and as a bedding medium for young plants. For reptile-keepers it is an ideal, sterile incubation medium for snake and lizard eggs.

VERNACULAR NAME Any everyday or local name of a TAXON, i.e., in any variety or language other than that of zoological nomenclature. For example, the northern viper *Vipera berus* is known by at least four different names in England alone: the common viper; the crossed viper; the adder; and the nadder. Such colloquial names have no standing or scientific importance in nomenclature.

VERNAL Pertaining to the late spring. *Compare* AESTIVAL; AUTUMNAL; HIBERNAL; PREVERNAL; SEROTINAL.

VERRUCA A wart or wart-like outgrowth on the skin.

VERRUCOSE Covered in warts.

VERTEBRA One of the bones of the spinal column (backbone) of a VERTEBRATE

VERTEBRAL Commonly used, when describing pattern or scaling, for the area covering the spinal column, e.g., any one of the scale rows along the mid-dorsal line on the bodies of lizards and snakes; also a LAMINA on the CARAPACE of a chelonian.

VERTEBRAL LINE A stripe running the length of the centre of the back, lying over the spinal, or vertebral, column.

VERTEBRAL PROCESS An elevated ridge running the length of the centre of the back of certain species of snake, e.g., *Crotalus durissus*, in which it is especially noticeable.

VERTEBRATA *See* VERTEBRATE.

VERTEBRATE An animal characterized by a bony or cartilaginous skeleton, a well-developed brain, and a backbone or vertebral column consisting of a series of ring-like structures (vertebrae) which protects the spinal chord. The group contains mammals, birds, fishes, reptiles and amphibians, each of which form a class and which together form the subphylum Vertebrata within the PHYLUM Chordata.

VERTICAL Describing the pupil in certain frogs, lizards and snakes which, in bright light, has an elliptical, slit-like opening with its long axis running vertically from top to bottom of the eye.

VERTICAL REPLACEMENT The manner of TOOTH REPLACEMENT in which a new tooth or succession of teeth is situated in a vertical row at the base of the acting teeth, progressing

quickly into a socket once it becomes empty.

VERTICIL A circular arrangement of parts around an axis; a whorl.

VERTICILLATE Arranged in, or having, verticils (i.e., whorls), and normally used to describe the position of the scales on the tail of certain lizards, such as the spiny-tailed iguana (*Ctenosaura*).

VESICULATION The development of blisters.

VESTIGIAL Relating to a vestige, or remnant, of a member or organ which at one time was more fully developed and which, during the evolution of the species, has become reduced in size and function, e.g., the vestigial PELVIC GIRDLE of a snake.

VESTIGIAL ORGAN *See* VESTIGIAL.

VIABLE Capable of normal growth and development; capable of living, as in viable eggs.

VILLIFORM; VILLOSE Resembling a VILLUS.

VILLUS (pl.) **VILLI** Any one of the many finger- or hair-like projections of the mucous membrane lining the small intestine of many vertebrates. There are also villi in the CLOACA of certain male urodeles and coating the tongues of many lizards.

VIPERID A snake of the family VIPERIDAE, possessing enlarged hinged fangs at the front of the upper jaw.

VIPERIDAE Vipers. Family of the SQUAMATA, suborder SERPENTES, inhabiting most of the world except the Australasian region.

Over 130 species in 11 genera in three subfamilies.

VIPERINE Of, relating to, or resembling a viper, but also descriptive of any snake having an angular, arrow-shaped head.

VIRUS INFECTION Any one of several diseases caused by one of a group of sub-microscopic intracellular bodies, many of which are PARTHENOGENIC, capable of reproduction only within the cells of other organisms, including reptiles and amphibians. Normally, virus infections show common infection symptoms such as listlessness and lack of appetite, or abnormal skin growths and tumours. Secondary bacterial infections, such as *Aeromonas*, then frequently bring about changes that are necessary for the continued development of the disease.

VISCERA The soft, large internal organs of the body; the entrails, particularly those in the abdominal cavity (COELOM). Post-mortem examination of amphibians and reptiles will often expose a cause of death which would not have been apparent from external examination alone.

VISCERAL GOUT *See* GOUT.

VISCOUS Sticky and thick; viscid.

VITELLOGENESIS The production of egg YOLK in the FOLLICLES begun when oestrogen stimulates the liver to commence converting lipids from the body's store of fat generating VITELLOGENESIS

VITELLOGENIC ACTIVITY The growth and deposition of YOLK.

VITELLOGENIN A protein, manufactured in the liver after oestrogen stimulation, that is the foundation to several yolk proteins. The growing follicles absorb vitellogenin from the bloodstream.

VITAMIN DUSTING *See* SUPPLEMENT DUSTING.

VITTA A band or stripe of colour. Generally used for the dark stripe running from either the snout or the eye along the side of the head to the tympanum or shoulder in many anurans, e.g., HYLIDAE. When a vitta is particularly broad it is often termed MASK

VIVARIUM (pl.) **VIVARIA** A tank, cage or enclosure in which live animals (generally reptiles and/or amphibians) are kept under sterile or natural conditions for research, study etc., or a building containing a collection of such tanks and enclosures.

VIVIPARITY A kind of reproduction in animals in which the embryo develops within the body of the adult female and receives its nourishment directly via some form of placenta, resulting in the eventual birth of live young. Viviparity occurs in many reptiles and amphibians, as well as in some invertebrates, certain fishes, and most mammals, all of which are termed 'viviparous'. *Compare* OVIPARITY; OVOVIVIPARITY.

VIVIPAROUS *See* VIVIPARITY.

VOCALIZATION Vocal intercourse within, and between, species, the most well-known examples being the noisy breeding calls of frogs and toads. Certain lizards can also vocalize and in some, such as the Asian Tokay gecko *Gekko gecko*, the call is so loud it can be heard over a considerable distance.

VOCAL POUCH The VOCAL SAC of frogs and toads.

VOCAL SAC An elastic, thin-walled, inflatable pouch on the throat, or at the sides of the neck, in male frogs and toads. In the majority of species there is a single sac lying beneath the throat or chin, whilst in others it is paired, swelling out between the TYMPANUM and shoulder. During VOCALIZATION the sac is filled with air from the lungs and acts as a resonating chamber. A few species have thoracic and abdominal vocal sacs.

VOLAR Of, or relating to, the sole of the foot or the palm of the hand.

VOLUNTARY MAXIMUM The highest possible temperature that a POIKILOTHERM can withstand before it is forced to seek shade or move below ground to AESTIVATE. *See* CRITICAL MAXIMUM; OPTIMUM TEMPERATURE RANGE.

VOLUNTARY MINIMUM The lowest possible temperature that a POIKILOTHERM can withstand before it is forced to seek the sun's rays, either direct or diffused, in order to bask, or to move below ground in order to HIBERNATE. *See* CRITICAL MINIMUM; OPTIMUM TEMPERATURE RANGE.

VOMER The thin, flat bone in the roof of the mouth, located directly behind the PREMAXILLA, that forms part of the palate.

VOMERINE TEETH Short, conic projections lying on the VOMER, in the palate of amphibians close to the internal nasal openings. Occurring in many species, vomerine teeth are often a useful guide to their identification.

VOMERONASAL ORGAN *See* JACOBSON'S ORGAN.

VON LINNÉ *See* LINNAEUS.

W

WAGLER, Johann Georg,(1800-1832), German zoologist and medical practitioner whose several herpetological works include *Naturliches System der Amphibien* (1830). Author of several taxa.

WAGLERIN A component of PIT VIPER venom which blocks nerve impulses to muscles causing paralysis.

WALLACE, Alfred Russel, (1823-1913), English naturalist and a contemporary of Charles Darwin. As a result of his observations in the East Indies Wallace came independently to the theory of evolution by NATURAL SELECTION and both men presented their ideas in a joint paper published by the Linnean Society of London on 1 July 1858.

WALLACE LINE The imaginary boundary passing between Bali and Lombok, Borneo and Celebes, and east of the Philippines separating the Asiatic and Australasian fauna. The line was named after Alfred Russel Wallace, English naturalist and traveller who, with Darwin, co-founded the theory of evolution.

WALTZ The 'dance' participated in by male and female *Ambystoma* as they encircle each other whilst attempting to engage their snouts with each other's CLOACA. *See* TAIL WALK.

WARM BLOODED *See* ENDOTHERM; HOMOIOTHERM.

WARNING ATTITUDE *See* UNKEN REFLEX.

WARNING CALL *See* MALE RELEASE CALL.

WARNING COLORATION The obvious or showy marking or colouring by which an animal advertises to potential attackers that it is dangerous or noxious. Many venomous snakes have conspicuous markings, as do poisonous and foul-tasting amphibians. *See* APOSEMATIC

WARNING CROAK *See* MALE RELEASE CALL.

WARNING DISPLAY The display in cobras and related species involving neck spreading and usually accompanied by hissing and an erect strike posture. Similar warning displays are seen in certain non-venomous species also, some of which can be classed as MIMICS.

WART Any well-defined, cornified raised area on the skin of certain amphibians, particularly toads of the *Bufo* genus, of which they are characteristic. Warts may also occur on the skins of reptiles from time to time particularly lizards and these usually signify a viral infection or, in the case of captive specimens, a lack of exposure to ultra-violet light.

WASHINGTON CONVENTION *See* CITES.

WATCHGLASS An alternative term for the SPECTACLE in snakes and some lizards.

WATERDOG A totally aquatic salamander of the family

221

PROTEIDAE represented by three species in the genus *Necturus*, and occurring in North America.

WATER MEADOW An area of low-lying grassland that borders a river and which is regularly flooded. Improved river management is leading to the rapid disappearance of this special type of habitat which, at present, is important to many species of amphibian and reptile.

WATER MOCCASIN *See* MOCCASIN.

WATERSPOT A physical irregularity in the calcification of a reptile egg shell, appearing as a slightly elevated star-shaped mark, not unlike a snowflake crystal, on the surface.

WATTLE A loose fold of skin hanging from the neck or throat of certain lizards and employed by males during courtship, when it can often be erected or extended by means of the HYOID BONE and used in display.

WAXWORM The larva of the waxmoth *Galleria melonella*, a member of the Lepidoptera, family Pyralidae. Commonly called tebos, waxworms are cultured as a nutritious food item for many reptiles and amphibians.

WCI Acronym denoting Wildlife Conservation International (New York Zoological Society).

WCMC Acronym denoting the Wildlife Conservation and Management Committee (AAZPA).

WEB The thin sheet of skin joining the digits of many amphibians and reptiles.

WEDGED PREOCULAR In snakes of the genus *Coluber* (the racers and whipsnakes), the condition in which the lower PREOCULAR projects downward between the adjoining LABIAL scales.

WERNER, **Franz**, (1867-1939), Austrian herpetologist who published many papers, particularly on snake systematics (1912, 1924, 1929), and was responsible for the much admired revision of the reptiles and amphibians in Brehm's *Tierleben* (volume 4).

WETLAND Any area of marsh or fresh water, including canals, streams and rivers et., which is home to many types of reptiles and amphibians. Much attention and effort has been directed at the conservation of this kind of habitat which is under constant threat from drainage and reclamation.

WETTSTEIN-WESTERHEIMB, **Otto von,** (1892-1967), Austrian herpetologist. Published *Herpetologica Aegaea* (1953) in which he resumed the research of Werner on the reptiles and amphibians of the Aegean.

WHIPSNAKE Any of several long, slender, fast-moving non-venomous snakes of the COLUBRID genus *Coluber,* such as *Coluber viridiflavus* of Europe. Also any of several other slender non-venomous snakes, such as *Masticophis flagellum* of North America.

WHITE ALBINO A fairly recently devised term sometimes used to describe SNOW varieties.

WHORL Term given to the scales, in

222

lizards, that surround a single portion of the tail in a symmetrical set, and used generally in reference to sets that are obviously quite different, as in the tail of the spiny-tailed iguanas (*Ctenosaura*), where rows of enlarged spiny scales alternate with several rows of small, flat, overlapping scales.

WEIGMANN, Arend Friedrich August, (1802-1841), German zoologist at the Berlin Natural History Museum. Compiled the first standard work on the HERPETOFAUNA of Mexico: *Herpetologia Mexicana* (1834).

WINDOW The area of transparent skin on the lower eyelid of certain lizards, particularly members of the SCINCIDAE, which enables the animals to see when the eyes are closed. In the ocellated skinks (*Ablepharus*) of south-eastern Europe and west Asia, the upper eyelid has fused firmly to the lower one, forming a snake-like SPECTACLE, or permanent window over the eye.

WING 1. The wide, delicate sheet of skin that stretches from the forelimb to the hind limb in the so-called 'flying dragon' (*Draco*). Supported by several long false ribs, the membrane is used for gliding between trees of up to 30 metres apart. At rest, the 'wings' are folded fan-like against the flanks. **2.** The sideways extension of separate sections of the chelonian PLASTRON which form the BRIDGE.

WING MEMBRANE *See* WING (sense 1).

WOLFFIAN DUCT Either of the pair of ducts found in amphibians, that carries URINE to the CLOACA from the kidney. In males, it has a duel purpose and also carries spermatozoa from the TESTES. It is replaced by the URETER in reptiles and is found only in males, where it forms the EPIDIDYMIS and VAS DEFERENS.

WOLTERSTORFF, Willy, (1864-1943), German herpetologist and zoologist. Produced many herpetological papers many of which were on urodeles in which he specialized. An important collection of this group was started by him at the Magdeburg Museum but was lost during World War 2.

WOMA *Aspidites ramsayi,* an Australian, nocturnal python related to the black headed python *A. melanocephalus* and inhabiting the arid regions of northern Australia.

WORM The pale, fleshy, worm-like lure on the tongue of the alligator snapping turtle *Macroclemys temmincki* used to entice prey into its open jaws.

WRIGHT, Albert Hazen, (1879-1970), American herpetologist, author of numerous significant papers on anuran distribution and development. In 1949 he co-authored *Handbook of frogs and toads of the United States and Canada* and later, in 1957, the two-volume *Handbook of snakes of the United States and Canada,* with his wife Anna Allen Wright.

WWF Acronym denoting the World Wide Fund for Nature. Formerly known as the World Wildlife Fund, this global organization is one of the world's largest and most respected independent conservation institutions. Active in over 90 countries, it works continually to stop the progress of the rapidly increasing destruction of the natural world.

X

XANTHISM; XANTHOCHROISM
An ABERRANT condition involving
the presence of an excessive quantity
of yellow pigment. A xanthic
specimen is a colour MORPH that
has yellow as its dominant colour.

XANTUSIIDAE Night lizards.
Family of the SQUAMATA,
suborder SAURIA, inhabiting
southern North America, Central
America and the West Indies. Some
12 species in four genera.

XENOPELTIDAE Sunbeam snake.
Family of the SQUAMATA,
suborder SERPENTES, infraorder
Henophidia, inhabiting South-East
Asia from Burma to southern China.
MONOTYPIC.

XENOSAURIDAE Knob-scale
lizards. Family of the SQUAMATA,
suborder SAURIA, inhabiting
Mexico and China. Four species in
two genera.

XERIC Relating to hot and dry
conditions, or to the animals living in
areas where those conditions occur.

XERIC PATTERN A type of
BREEDING PATTERN occurring in
certain types of frogs, which is
effected greatly by the outside
surroundings, the main influence
being rainfall. Species having a xeric
breeding pattern do not usually have
a fixed breeding season but instead
breed throughout the year in reaction
to rain.

XIPHI- A prefix meaning shaped
like, or resembling, a sword.

XIPHIPLASTRON (pl.)
XIPHIPLASTRA Either one of the
hindmost pair of bones in a
chelonian PLASTRON.

Y

YACARE An alternative spelling to JACARE.

YAWN The wide stretch of the jaws that precedes the start of ECDYSIS in snakes and lizards or which, in snakes, follows the swallowing of a large prey item to realign the jaws.

YELLOW BEARD *See* BARBA AMARILLA.

YOLK The spherical mass of normally yellow nutritious substance that maintains the developing embryo within the YOLK SAC.

YOLK DEPOSITION The deposit of YOLK into the OVUM.

YOLK PLATELET Any one of the somewhat large and flattened granules of yolk occurring in the eggs of amphibians.

YOLK SAC The membranous sac, linked to the ventral surface of the embryo, that contains the YOLK in reptiles. Upon hatching from the egg, the yolk sac is drawn into the abdomen and provides the young reptile with nourishment during its first few days of life until it can feed itself.

YOLK SYNTHESIS The development of yolk, originally beginning in the liver and finishing in the OVUM.

226

Z

ZIGZAG FORM Descriptive of variable patterned snakes that frequently have their dorsal spots or blotches fused together to form zigzag-like patterns.

ZONARY Term used to describe the concentric pigmented areas occurring on the SCUTE of a chelonian shell.

ZONURE A lizard of the genus *Cordylus*, family CORDYLIDAE, inhabiting the hot, rocky regions of tropical Africa

ZOOGEOGRAPHICAL REGION Any one of the six geographical divisions of the world, sometimes called 'faunal regions', contrived in conformity with the distribution of land animals. The regions comprise PALAEARCTIC, NEARCTIC, NEOTROPICAL, ORIENTAL, ETHIOPIAN (or Afrotropical) and AUSTRALASIAN, and each has its own unique collection of animal species.

ZOOGEOGRAPHY A branch of zoology concerned with the geographical distribution of animals and the primary causes of this.

ZOOGEOGRAPHICAL RECORD An important and comprehensive annual publication which includes precise sources of reference from the literature of zoology with separate sections for 'Amphibia' and 'Reptilia'. Details all new taxa validated during a specific year.

ZOOLOGICAL NAME In binomial nomenclature, the SCIENTIFIC NAME of an animal TAXON.

ZOOLOGICAL NOMENCLATURE The system of SCIENTIFIC NAMES for animal taxa and the stipulations for the creation, management and usage of those names.

ZOONOSIS (pl.) **ZOONOSES** A disease or infection that can be passed to man from lower vertebrates. Only very few reptilian diseases are transmittable to man.

ZOOPLANKTON The tiny, free-floating organisms that inhabit aquatic systems and upon which many amphibian larvae depend during their development.

ZOOTOXIN A poisonous secretion produced by an animal and, in herpetology, refers to the toxic skin secretions of amphibians and to snake venoms.

ZYGAPOPHYSIS (pl.) **ZYGAPOPHYSES; PROCESSUS ARTICULAR** Any one of the several processes on a vertebra that articulates with the corresponding process on an adjacent vertebra.

ZYGODACTYL; ZYGODACTYLOUS Having the toes in pairs, two facing forwards and two facing backwards, as in birds of the Picidae family (the woodpeckers), but also used in herpetology when relating to the feet of the CHAMAELEONIDAE. The fused toes of chameleons have been modified into claws for climbing and grasping: the fore limbs having three inner toes and two outer ones, and

the hind limbs having two inner toes
and three outer ones.

ZYGOTE A fertilized female
GAMETE, resulting from the union
of a SPERMATOZOON and an
OVUM, or the individual organism
resulting from such a cell.

♂ Symbol indicating male.

♂ ♂ Symbol indicating several males.

♀ Symbol indicating female.

♀ ♀ Symbol indicating several
females.

< Symbol indicating less than.

> Symbol indicating more than.

Printed and bound by CPI Group (UK) Ltd, Croydon, CR0 4YY

03/10/2024

01040333-0002